T0361614

DIMMING THE SUN

Also by Thomas Ramge,
published by The Experiment

Who's Afraid of AI?: Fear and Promise in the Age of Thinking Machines

The Global Economy as You've Never Seen It: 99 Ingenious Infographics That Put It All Together
(with Jan Schwochow)

DIMMING THE SUN

THE URGENT CASE FOR GEOENGINEERING

THOMAS RAMGE

NEW YORK

Dimming the Sun: *The Urgent Case for Geoengineering*

Originally published in Germany as *Die Sonne dimmen: Wie Geoengineering die Menschheit vor der Klimakatastrophe retten kann* by Penguin Random House Verlagsgruppe GmbH in 2024. First published in North America in revised form by The Experiment, LLC, in 2025.

The Experiment, LLC
220 East 23rd Street, Suite 600
New York, NY 10010-4658
theexperimentpublishing.com

Library of Congress Cataloging-in-Publication Data available upon request

ISBN 979-8-89303-054-9
Ebook ISBN 979-8-89303-055-6

Jacket and text design by Beth Bugler
Cover photographs by Adobe Stock/Vector Light Studio (Sun) and matis75 (Earth)
Illustrations by Peter Palm
Author photograph by Michael Hudler
Translation by Monika Werner

Manufactured in the United States of America

First printing March 2025
10 9 8 7 6 5 4 3 2 1

For Moritz

CONTENTS

INTRODUCTION:
Sulfur in the Sky

Not So Cool Cooling Credits

Luke Iseman and Andrew Song stand in a parking lot in the San Francisco Bay. It's a sunny April morning. It's still cool, but there's hardly any wind. A large helium bottle lies on the asphalt next to their RV. Iseman guides the hose from the bottle into the opening of a weather balloon. Song opens the valve, and the white balloon slowly fills to a diameter of around seven feet. Iseman keeps a tight grip on it. He disconnects the hose and closes the balloon opening with black duct tape. "We're definitely not scientists," he says, a man in his mid-thirties with a Mohawk, while he smiles mischievously at the CBS television crew. Then he lets go. The balloon rises surprisingly slowly at first but then quickly picks up speed for its journey into the stratosphere.

The two founders of the California start-up Make Sunsets look upward with satisfaction. Sulfur dioxide is mixed into the balloon's helium. Iseman flashes his mischievous grin again: "It's pathetic but it's a start." We've heard this story before: California founders who want to change the world and get rich in the process. Make Sunsets's story is not about the next "disruptive" app—for, say, better car sharing, cheaper money transfers, or more comfortable shopping—that one can choose to use or not. Their business plan is based on a technology that will affect—if put into use—every single human, animal, and plant on the planet. The stakes could barely be higher. The scheme they're testing out is simple and cheap. It could become a stopgap solution that buys humanity the time to transition more smoothly to a post-fossil age. Or, conversely, it could slow down decarbonization, stir up dangerous political conflict, or even kick off a new ice age.

Luke Iseman and Andrew Song met in 2015. At the time, Iseman worked as Director of Hardware at the renowned start-up company Y Combinator. Song worked as an Outreach Manager for Indiegogo, a financing platform for entrepreneurs to raise funds for their projects. In 2022, they convinced investors to commit 750,000 dollars to their idea. Their weather balloons will burst at an altitude of around twelve miles, and release sulfur dioxide into the stratosphere. They don't just make money with this; they also help the climate. At least, that's how the Make Sunsets entrepreneurs see it. The young company makes its customers a seemingly attractive offer. For ten dollars, they use weather balloons to release one gram of sulfur dioxide (SO_2) into the stratosphere, where the sulfur gas combines with tiny water droplets to form a white layer of mist that lasts for at least a year. This layer reflects solar radiation back into space, reducing the penetrative solar energy that heats the Earth. One gram of sulfur dioxide in the stratosphere offsets the climate damage of one ton of carbon dioxide, claims Make Sunsets. Iseman and Song regularly report on their company website how many kilograms of SO_2 they are sending into the stratosphere.[1] Skeptics consider the start-up to be a PR stunt. The amount of sulfur is far too small to have a cooling effect in the stratosphere. Nevertheless, the PR effect itself is measurable. Thanks to the media, Iseman and Song are broadcasting their central message around the world: Planet Earth can be saved from climate collapse only if people dim the sun as quickly and as effectively as possible. Climate researchers and activists throw their hands up in horror.[2]

For around fifteen years, the research community debated whether so-called solar geoengineering should even be researched in more detail. The Intergovernmental Panel on Climate Change (IPCC) consistently ignores the issue, partly because several countries are blocking its discussion. Most IPCC scientists and climate activists from around the world are keen to avoid the impression that there is a simple, quick, and cheap solution to the greenhouse effect. And then, a small start-up from California comes along and, without any serious scientific support, turns solar radiation modification (SRM) into a questionable business model.

Above all, Make Sunsets shows how geoengineering should *not* be done. Dimming the sun—potentially one of the most helpful or most dangerous interventions in the earth system that humans can make—is of course not to be driven by market mechanisms and companies making a quick buck on flimsy certificates. A decision of solar geoengineering obviously needs to be based on as deep a scientific understanding as possible. And there can be no doubt that cooling the planet must be subjected to strict international regulation backed by a vast majority of countries. Still, Luke Iseman and Andrew Song might get some credit for a constructive role in a complex situation—for seeing things earlier than others. Sometimes, it takes breaking taboos to force open deadlocked debates and give politics new impetus. The climate debate urgently needs such an impulse. It is stuck in a constant loop in parliaments, international forums, and talk shows. Meanwhile, temperatures continue to rise—as do emissions. Solar geoengineering will disrupt discussions and negotiations about climate policy in the coming years and steer them in a new direction. And the technology will make its way into the world at the latest in the next decade—whether we want it or not. And this is why I wrote this book.

The debate on solar geoengineering is slowly gaining traction, but it is a recent development. I stumbled upon the subject about ten years ago. And like most experts with whom I've discussed dimming the sun since, my first impulse was: This is crazy. Not crazy in the daring sense, but, rather, reckless. But, with the ever-growing doubts on the speed of decarbonization and the statistical chances of achieving the aims of the Paris Agreement, I started to take a closer look at this crazy idea about five years ago. Solar geoengineering still looks absurd to me, but so does the current path of climate policy. And considering rising temperatures and extreme weather events, I came to this conclusion: It would be crazy to *not* prepare for a world in which dimming the sun in the most responsible way possible might become humanity's best choice in dire circumstances.

To avoid any misunderstandings from the get-go: I am not arguing to start geoengineering the atmosphere as soon as possible, and within my years of research, I have not come across a reasonable scientist in the field who does. But, like many of these scientists, I am troubled by the lack of open consideration on the topic. Given the facts we know today, and considering the trends we see, I am convinced that solar geoengineering bears great potential to save the world from the worst effects of climate change.

Conversely, it is crystal clear that human intervention as profound as technologically cooling down the planet bears great risks. I am worried about these risks as well, but what worries me much more is that humanity is not looking into these risks as carefully as possible, as *soon* as possible— and that we might one day dim the sun recklessly. This is why this book took off with the balloons of Make Sunsets. They foreshadow the climate world of tomorrow.

It is no longer a question of whether solar geoengineering will be carried out, but only when, whether planned and in cooperation with critical geopolitical actors or by rogue agents. Dimming the sun is cheap. One could also say: too cheap. Maintaining the Earth's temperature at current levels using sulfur aerosols is expected to cost less than twenty billion dollars annually. This would save hundreds of billions in climate change costs in the short term and probably trillions in the medium term. Rich states or an alliance of smaller countries from the Global South that suffer particularly severely from climate change would be able to dim the sun. A tech-fanatic tycoon with the personality of Elon Musk might feel called upon to turn down the Earth's thermostat. A radical non-profit organization with a wide-reaching enthusiastic audience could also announce that the suffering caused by climate impacts is too great—and then take independent action. Solar radiation modification may not be a perfect solution, but it is better than none at all. And there will be a country that would permit balloons or airplanes to take off. Global geoengineering chaos is conceivable: different actors sowing clouds and dimming the sun, uncoordinated and using various methods on several continents.

DARPA, the research arm of the US military, set up a program in 2022 to uncover so-called rogue geoengineering.[3] Clearly, the US wants to know if someone is messing around with the climate somewhere in the world. How will the US respond if India attempts to lower temperatures and increase rainfall in South Asia in 2035 after multiple heat waves with millions of heat-related deaths? And how could the Chinese government react to such an intervention if it leads to even more periods of heat and drought in Southern China? Going it alone in geoengineering will create serious geopolitical conflicts. Conversely, everyone around the world has a common interest in avoiding climate collapse. Climate protection is one of the few policy areas in which the different camps of the new, multipolar world order still cooperate. In an optimistic scenario, a new decision-making mechanism for solar geoengineering could lead to greater international cooperation again.

Methadone for a Carbon-Addicted Humanity

"Geoengineering" (or climate engineering as many scientists prefer) is an umbrella term. It refers to two different approaches to manipulate the climate on a planetary scale: carbon dioxide removal (CDR) and solar radiation modification (SRM). Although CDR also faces some criticism—i.e., providing an excuse for not cutting emissions faster—removing CO_2 from the atmosphere has made it into the mainstream of global climate policy. Most experts consider CDR an expensive but necessary technique to achieve a net-zero economy, to compensate for the emissions that are hard to avoid, and I fully agree. In chapter 2, I will give a brief overview of the most common approaches to how CO_2 can be extracted and stored away for good, albeit at high financial cost. However, in this book, I will focus on discussing the controversial side of geoengineering, solar radiation modification, for mainly three reasons.

First, as emissions and temperatures keep on rising, it is high time to put all options on the table, especially those that are not much talked about. Secondly, and surprisingly, when geoengineering information is presented as the last resort of climate policy, it increases the desire for decarbonization. And thirdly, and in contrast to all other climate protection measures including CDR, solar radiation modification is technologically feasible and works quickly.

The best indication for this is massive volcanic eruptions such as the one in the Philippines in 1991, which caused the global average temperature to fall by around 0.5°C (0.9°F) the following year. Artificially darkening the sky with sulfur is tempting today, because we already live in a world with around 1.5°C (2.7°F) warming above preindustrial average temperatures, the average from 1850 to 1900. But how attractive will dimming the sun be when the Paris Agreement's upper limit of "well below 2°C (3.6°F)" is ancient history and the Earth system is heading toward a warming of 3°C (5.4°F)? Because then the so-called tipping points in the climate system threaten to fall like dominoes: First, large parts of the (by then non-existent) "eternal ice" at the poles melt. Then, the permafrost in the High North thaws and releases enormous amounts of additional greenhouse gases. What happens if the monsoons fail, and the number of hurricanes increases dramatically? And what happens if the Atlantic Meridional Overturning Circulation (AMOC) and the Gulf Stream collapse with it, as feared by a 2024 Dutch study that drew worldwide attention?[4] What political mechanisms are set in motion if island states sink into the sea and Bangladesh, which is at high flood risk, is mostly flooded? What if parts of Central Africa and South Asia become uninhabitable and migration pressure on the Northern

Hemisphere steadily increases? What if the death toll from flood damage continues to rise and food security in the Global South declines even further? Perhaps then the "Climate Endgame"—humanity's fight for survival—will begin, as prolific risk researcher Luke Kemp from the Notre Dame Institute for Advanced Studies (NDIAS) fears.[5] Unfortunately, this horror scenario is not a pessimistic thought experiment only for fans of dystopian novels. If the current climate policy of the Paris Agreement continues, the Earth is expected to warm up by 2.5 to 2.9°C (4.5 to 5.2°F) by the end of the century.[6] However, this assumes that there are no setbacks in decarbonization, for example, when populist governments withdraw from the agreement. According to the UNEP Emissions Gap Report, greenhouse gas emissions would have to fall by at least 28 percent in the next five years to maintain the chance of limiting global warming to at least 2°C (3.6°F).[7] Maybe a miracle will happen, and the world's major countries will pull themselves together and reduce CO_2 emissions faster and more radically than announced. But how likely is that? 2023 had a double negative record. It was the hottest year in around 120,000 years.[8] 2024 set new records again. Nevertheless, humanity burned more fossil fuels than ever before. Of course, everyone responsible for climate policy knows this. But almost everyone acts as if we can still achieve the Paris climate goals if humanity just knuckles down on decarbonization. Europe enjoys the pioneering role here while notoriously overestimating its contribution to global climate protection. Is this demonstrative calculated optimism strangely naive, or cynical and defeatist? Or is it simply stupid and irresponsible toward future generations?

In the coming decade, the demand for solar geoengineering will be part of the political mainstream in many countries around the world. Sooner or later, a government, a coalition of states, or a non-state actor will rush forward and cloud the stratosphere. Therefore, it is high time for science and the public, politics, and international climate bodies to systematically clarify the three central questions of solar geoengineering.

1. What technical approaches could be used to dim the sun effectively and safely?
2. What risks would solar geoengineering pose, and what side effects would it entail? Particularly important in this context are the geophysical uncertainties and, above all, the political and social concerns.
3. How could the (vast majority of) humanity agree on an effective decision-making mechanism for or against deployment, and what legal framework do we need for this?

This book provides initial answers to these questions. These answers will face criticism. I will expose myself to the accusation of advocating a megalomaniacal pseudo-solution that undermines current climate policy. I look forward to the debate, provided it happens based on scientific and economic arguments. In anticipation of the discussions, I would ask you to note the following information, which should be included on every package insert for solar geoengineering:

Dimming the sun is the methadone program of a CO_2-addicted humanity. It is cheap and works quickly, but unfortunately, it does not offer a long-term solution. Solar geoengineering does not address the causes of climate change. Instead, there is a risk of several adverse rebound effects. It would be great if humanity could forgo solar geoengineering and get climate change under control before the world goes off the rails. But I no longer believe in this possibility. In a realistic scenario, shading the Earth can provide us a transition period to help us to finally move away from carbon while simultaneously mitigating the dire climate impacts during the withdrawal phase. No more and no less. In the long term, we, the carbon junkies, have to beat our addiction. How can this balancing act succeed?

Powerful Technology and Human Flaws

As chapter 2 will show in detail, injecting aerosols into high layers of air (so-called stratospheric aerosol injection, or SAI for short), creating a thin, white veil around the globe, seems to be the most technically promising approach. However, we could also artificially form more clouds in lower altitudes over the sea, or disperse so-called cirrus clouds over a large area to allow more heat to escape from the greenhouse Earth, as they hinder the heat radiation back into space. Solar sails in space are also theoretically conceivable, with which solar radiation could be reduced as required, possibly regionally differentiated. Some tech-visionaries think we might even be able fog the Earth with moon dust.

The greatest risk of solar geoengineering lies not in the technology but in our human flaws: our capacity for self-deception and penchant for easy excuses. "We have a technical solution," we might say, "so we don't have to change our behavior." This is what Newt Gingrich, the former Republican Speaker in the US House of Representatives, argued in the 2000s.[9] Climate economists call this danger the "moral hazard of geoengineering." From a game theory perspective, there is a heightened form of the free-rider effect here. Actors could benefit from geoengineering without being involved and, at the same time, continue to produce CO_2 without personally feeling the consequences of their harmful behavior. Dimming the sun as an interim measure makes sense only if all (important) parties involved understand that an interim solution is exactly that; an incentive to use the time gained with all their might to achieve an actual solution.[10] If you are in a boat with a leak, it's not enough to provisionally plug the hole if you want to sail around the world.

Conventional solutions for dilemmas are, for the most part, ineffective. A pragmatic solution to reduce the "moral hazard" could look like this: As many actors as possible commit to an international agreement to understand and use geoengineering exclusively as a way to gain more time for the decarbonization process. They agree to limit the technical intervention to, for example, 2°C (3.6°F) of warming (so-called peak shaving) or to dynamically reduce the additional warming by 50 percent. At the same time, the "United Geoengineering Nations" commits to driving green energies forward faster and more consistently—which will be easier in 2040 than today, when green energy will be cheaper than fossil energy almost everywhere in the world. Ideally, energy is then "too cheap to meter"—i.e., so cheap that it is no longer worth billing for it. This is not a technical utopia. It is already possible in sunny regions today. The large solar power plants on the Arabian Peninsula produce electricity for around one cent per kilowatt hour. The electricity meter is no longer worth the trouble.[11] With an abundance of green energy, there is an opportunity to gradually reduce the proportion of CO_2 in the atmosphere using energy-hungry extraction processes, carbon dioxide removal (CDR), to at least the current level of 420 ppm (parts per million). Within several decades, the carbon dioxide concentration could perhaps be reduced back to its preindustrial level of around 270 ppm. This would restore the previous balancing act of energy input through sunlight with heat radiating from the atmosphere into space (through infrared waves).

What does this mean in a nutshell? It is likely that humanity, based on current development paths, will soon live in a world more than 2°C (3.6°F) above preindustrial levels. A sad but plausible scenario is that our children and grandchildren will have to live with 3°C (5.4°F) of warming. With worsening climate impacts and, subsequently, greater suffering, there is an increased likelihood of erratic, uncontrolled use of solar geoengineering. Those responsible for climate change in politics and those knowledgeable about climate change in science must break the well-intentioned vow of silence regarding a powerful technology, even if or precisely because, like all powerful technologies, it is a double-edged sword. Whether it does more good than harm depends on who uses it, how, and for what purpose. Solar geoengineering will probably be carried out systematically in the next decade, not merely as a PR stunt by a start-up like Make Sunsets. The goal must be to dim the sun in a scientifically competent and politically responsible manner. Solar geoengineering is not too good to be true. The method is not good. But the truth is that humanity can use it to buy the time necessary to make the transition to a post–fossil fuel world.

1

CLIMATE:
Why We Have to Dim the Sun

Horror Temperature

The term "cooling limit temperature" sounds extremely technical. Physically, it describes the lowest temperature reached through evaporation at a certain humidity level. Biologically, this value is more important than the mere air temperature for us human beings. It is a question of survival. Our bodies cope much better at 38°C (100.4°F) in dry air than 30°C (86°F) with very high humidity. As long as there is low humidity, physics helps our biology. When it's hot, we sweat and go into the shade, hoping for a cooling wind, or we turn on a fan. The evaporation of sweat significantly lowers our body temperature. Sweat cannot evaporate if the heat has high humidity. A classic thermometer and a damp cloth can measure the cooling limit temperature. The cloth (and the moisture in it) is wrapped around the glass bulb at the bottom of the thermometer. At 100 percent humidity, the air and cooling limit temperatures are the same. The drier the air, the lower the temperature drops, thanks to the possibility of evaporation. At a rather low 30 percent relative humidity and 45°C (113°F) air temperature, the cooling limit temperature is "only" 28°C (82.4°F)—again, that's how far, the "limit" to which the thermometer can drop thanks to evaporation. In comparison, at 90 percent water vapor, it can drop just a couple degrees, from 45 to around 43°C (109.4°F).

We generally underestimate the effect. To us, sweating is merely unpleasant. We hear the air temperature value on the radio and know what to expect when we go outside. This will be more complicated in the future because the cooling limit temperature—called "wet-bulb temperature" after the classic measuring method—will more often determine our health and well-being. For millions, it will be the most important parameter for deciding about climate migration or the threat of death from heat. Unfortunately, this is not scaremongering; instead, the "wet-bulb effect" is a relatively simple calculation from a biological and climate-science point of view.

A healthy adult can survive in the shade for just six hours at a wet-bulb temperature of 35°C (95°F)—that is the air temperature minus cooling by evaporation. At this temperature, the skin can hardly release any heat into the air, no matter how much you sweat. The heart pumps more and more blood through the body, which is supposed to release heat through the skin, but the wet-bulb temperature makes this impossible. Drinking more is no longer of any use. The body still dehydrates. The temperatures inside the body ultimately rise to a deadly 42 to 43°C (107.6 to 109.4°F).[1] In the past, we only very rarely reached the threshold of 35°C (95°F) cooling limit temperature. When we did, it was usually in sparsely populated places such as the Indian-Pakistani border, the Persian Gulf, and the Gulf of Mexico.[2] Meteorologists do not always register wet-bulb extreme weather. Still, according to calculations by climate scientist Colin Raymond, one thing is for sure: In the last forty years, the cooling limit temperature has approached the deadly threshold more than three times as often than in the previous four decades.[3] That is not a weather coincidence but follows a physical principle that is also relatively easy to understand.

Warm air can hold more water vapor than colder air. Global warming is also causing more severe and longer-lasting droughts. However, the wet-bulb effect gets a double boost in hot and humid regions due to the higher relative humidity when temperatures rise. The British Meteorological Office has modeled the climate consequences in severe warming scenarios: With a global average of 2°C (3.6°F) of warming, large parts of the Indian subcontinent will regularly experience deadly wet-bulb temperatures. If global warming exceeds 2.5°C (4.5°F), this will likely apply to the entire tropics—in some regions, for several months a year.[4] The dangerous limit will also be reached regularly in the southern states of the US, Eastern China, and Brazil.

The climate novel *The Ministry for the Future* by Kim Stanley Robinson, written in the years following the Paris Climate Agreement of 2015, begins with a cruel scene. It takes place in the very near future, in the year 2025. Wet-bulb extreme weather occurs in the Indian state of Uttar Pradesh. The power grid collapses under the load of air conditioning. Some village residents first flee to the few houses where diesel emergency power generators are still running. The majority try to cool off in the local lake. When the water temperature also rises above 35°C (95°F), they all die from heat. This heat wave costs twenty million people their lives. (I won't reveal who survives at this point. *The Ministry for the Future* is exciting right to the last page.)

Dystopian literature plays out horror scenarios. The more plausible and, despite all the fictional exaggeration, the more imaginable they are, the stronger their effect. Robinson's story is based on scientific knowledge. And our scientific knowledge is based on the increase in cooling limit temperatures; the current climate models predict with a high degree of certainty that many hot, humid regions will remain habitable from the middle of this century only with radical adaptation measures. Air conditioning for everyone is, by then, the minimum standard. People can go outside briefly—or with cooling suits.

The World with 3°C (5.4°F) of Warming

What would a world with 3°C (5.4°F) of warming look like by the end of this century? We can't know for sure. Future forecasts are based on a linear continuation of past developments. Climate change and its impact on the Earth system are proving to be non-linear in many ways. The higher the temperatures climb, the more unpredictable the effects. All signs today indicate that the negative consequences will not "just" increase linearly; that 3°C (5.4°F) of warming will cause twice as much damage as the previous 1.5°C (2.7°F). They will cause *many* times more costs to finances and human suffering. We can only guess at the dimensions to which this extends. Science has no more than a vague idea what effects will be triggered by increasing warming and when specific tipping points could be reached in the climate system. But if the stability of just one of the critical climate elements, such as the ice at the poles, topples this will (very likely) trigger cascades of dangerous knock-on effects. Science does not know the, by definition, "unknown unknowns" in the complex climate system. But a plausible scenario based on scientifically proven assumptions for a world with 3°C (5.4°F) plus looks something like this:

This 3°C-plus (5.4°F) is an average value, as temperatures over the oceans are rising more slowly than on the continents. The increase will be closer to 4 or 5°C (7.2 or 9°F) in many regions, including continental Europe, Central Asia, and interior North America. Generally, the following pattern applies: Where it is hot, it gets hotter. Where it is dry, it gets drier. Where it rains, it rains more often and more heavily.

This means there are many extreme heat waves and periods of drought in areas that are already vulnerable to them. These include, among others, California, the Midwest, and Central America, the northern part of Africa, Southern and Eastern Europe, and, of course, the Near and Middle East, large parts of Central Asia, South China, and Australia. Forest fires will not only increase the heat, but they will also severely pollute the air. Rationing of drinking water will be part of everyday local politics, even in rich countries. Around a quarter of the world's population will have to live in extreme drought for at least one month of the year. Depending on the success of adaptation measures, many people will die prematurely every year due to heat stress. Children and older people are particularly at risk, especially in cities, where the heat builds up dramatically during the day, and there is less cooling at night than in rural regions. Mosquitoes will migrate north, and in the Southern Hemisphere, they will migrate south, and transmit diseases. Crop failures will increase poverty and hunger, especially in those countries where agriculture accounts for a high proportion of their gross national product. Despite the heat and drought, the dry regions will also increasingly have to contend with damage caused by heavy rain since warm air, as described above, can store more water and then discharge it in bursts more often.

Along the equator, cyclones, hurricanes, and flood disasters will become more frequent and severe all year round, with the already known consequences for transportation infrastructure, power grids, water supplies, and wastewater. The likelihood of so-called compounding disasters—i.e., storm, flood, and heat disasters in quick succession, which will increase the suffering of the people affected, is also intensifying significantly. A cyclone destroys the power lines. A heat wave will follow, possibly with values close to wet-bulb temperatures. If we're lucky, people will have built a more resilient infrastructure and more robust houses by then, and perhaps in hot, humid areas, most residents will have emergency power generators for air conditioning. But is that likely? It is foreseeable that in a world with a temperature increase of 3°C (5.4°F) plus, no one will be able to insure themselves against natural disasters. The risk insurance business model will have become mathematically obsolete.

From today's perspective, it is scientifically certain: In a world with 3°C (5.4°F) of warming, there will be significantly fewer animal and plant species. How many will become extinct also depends on how quickly temperatures increase and change the weather patterns of the last ten thousand years. The faster warming progresses, the fewer the species that will be able to escape to higher or cooler habitats. Marine fish are already migrating to colder waters toward the polar ice caps. According to studies by Bielefeld University, the risk of extinction of an animal species will increase by around 26 percent by the end of the century—this is already the highest level of species extinction since the annihilation of the dinosaurs.[5] According to a WWF study, it is feared that in particularly species-rich ecosystems, such as the Amazon, Madagascar or the Galapagos Islands, a quarter of the species are threatened by just 2°C (3.6°F) of warming.[6]

If we reach 3°C (5.4°F), the acidification of the seas due to increasing carbon dioxide levels in salt water will result in only remnants of the coral reefs. It is almost certain that some island states will no longer be found on the map, including large parts of the Maldives, parts of the Bermuda archipelago, and the Seychelles. Through images of emaciated polar bears on thin ice floes, we have internalized the idea that temperatures in the Arctic are rising. However, it appears to be significantly faster compared to the Earth's average than scientists previously assumed. Studies now assume a factor of four.[7] The sea ice at the poles, which expands in winter and melts again in summer, has been overall declining sharply for decades. This process begins slowly in the large ice sheets over Greenland and Antarctica. In a medium scenario with 3°C (5.4°F) of warming, sea levels would rise by 60 to 70 centimeters (23.6 to 27.5 inches) by the end of the century. This would put numerous coastal cities in extremely challenging situations, including Bangkok, New Orleans, the large Dutch cities of Amsterdam, Rotterdam, and The Hague, as well as, in Italy, of course, Venice, with its cultural treasures.[8] But this scenario is still optimistic.

Polar ice melting is subject to a self-reinforcing feedback effect. White ice reflects solar rays much better than dark blue water and black-brown landmass. Less ice mass means less reflection of solar energy into space and, as a result, an increase in temperature, ultimately further accelerating ice melting. Scientists call this mechanism the "ice-albedo feedback." Presumably, the ice cap of Western Antarctica would begin to melt in a climate of 2°C (3.6°F) above preindustrial times. Their collapse would raise sea levels by more than a meter and a half (4.9 feet), although not yet by the end of the century. At the same time, the Greenland ice sheet continues to melt. If there was a sea-level rise in the 2.5-meter (8.2-foot) ballpark, large cities in rich countries could probably just hold it together with dikes and protective walls. However, several hundred million coastal residents around the world would have to relocate to higher inland areas.[9] More frequent and stronger storm surges would have the added effect of increasing *further* storm surges in climates with already stronger storms and heavier rain. The last polar bears, born when the sea ice was still intact, will then die on land because the North Pole will be ice-free for many months of the year.

Tipping Point Cascades

However, it's not the most predictable effects but rather the big unknowns that are the most disturbing about the 3°C (5.4°F) scenario, especially the so-called tipping points of important climate elements.[10] A tipping point describes threshold values in a range of systems. With linear development, nothing happens for a long time. However, if the change in a system exceeds a specific value, the system shifts into a new state and cannot easily return to the old one. The tipping at the threshold can occur abruptly, comparable to a pencil that suddenly tips and falls when slowly pushed over the edge of a table. Alternatively, a slow process can be set in motion, although its course is also irreversible. It is likely that the tipping point for Arctic Sea ice has already been reached today due to albedo feedback and that the ice sheets will continue to shrink, even if we can stop climate change. In a 3°C-plus (5.4°F) world, the same will apply to the Greenland ice sheet on land and, as mentioned earlier, to parts of Antarctica. The constant rise in sea level is just one consequence, and from today's perspective, not even the most dangerous. Strong warming of the poles—as mentioned, significantly higher than the global average—will most likely trigger a cascade of radical climate changes, reinforcing each other through further feedback effects.

As the High North gets warmer, the permafrost soils of Alaska, Canada, and Russia are in danger of thawing. Gigantic amounts of greenhouse gases are stored beneath them, including an exceptionally high amount of methane, which is especially harmful to the climate. If the soil thaws, it will escape into the atmosphere and is expected to increase global warming by several tenths of a degree. According to calculations by the Potsdam Institute for Climate Impact Assessment, the best-case tipping point for this is 2.3°C (4.14°F), but this may have already been reached.[11]

The melting of the poles can also have severe consequences for the system of ocean currents, especially for the Atlantic Meridional Overturning Circulation (AMOC). This ocean current system transports warm water to the north on its surface, where it cools down, sinks, and flows south again in deeper layers. The Gulf Stream is part of this system and, among other things, gives Northwestern Europe its mild climate. If the strength of this water circulation were to reduce, it would have massive effects on the entire global climate. The Southern Hemisphere would warm up significantly, the monsoons could weaken, and the Sahel zone could dry out further. The available studies on when the tipping point for the AMOC, essentially a large circulation pump, will be reached haven't reached a consensus —perhaps it will be at 2°C (3.6°F), perhaps only at 4 or even 6°C (7.2 or 10.8°F) of warming. But early warning signals, accumulated over several years, indicate that it is closer to 2°C (3.6°F) than 4°C (7.2°F).[12] A team of researchers led by the Dutch oceanologist René van Westen modeled a simulation, the results of which caused great concern in February 2024. Van Westen initially looked for warning signals in the southern part of the Atlantic between Cape Town and Buenos Aires and simulated changes in salinity over the last two thousand years. A clear and worrying result of the simulation is that a sudden collapse of the AMOC is possible if the gradual slowdown caused by global warming continues. According to the study, it is only uncertain whether something will happen in a year or in a century.[13]

However, polar ice melting, permafrost thawing, and disruption of ocean currents could be just the beginning of the cascade. As a result, the extensive coniferous forests of the High North, which make up a third of the world's forest area, could become desertified, followed in turn by the rainforests of the South, especially the Amazon—all as a result of long-lasting droughts in the middle latitudes of the Southern Hemisphere. A large proportion of the Earth's glaciers have already passed their tipping points and triggered an albedo feedback effect. In a world with significantly increasing carbon dioxide concentrations in the atmosphere—caused, among other things, by melting permafrost soils—fewer bright stratocumulus clouds are forming to naturally cool the Earth with their shade. At above three times the current CO_2 concentration, they dissolve completely. This would be a colossal domino in the series of important climate elements. The Earth's temperature would then rise by several additional degrees.[14]

A 2021 study of three million computer simulations of a climate model found that almost one-third of those simulations resulted in domino effects, even if temperature rise was limited to a maximum of 2°C (3.6°F), which is the upper end of what the Paris Agreement strives to meet. What happens then in a 3°C (5.4°F) scenario? Does the effect of the tipped dominoes then add up to an existential threat, as the study's authors suspect?[15] A network model analysis by the Potsdam Climate Impact researchers also showed that even temporarily exceeding the climate targets of the Paris Agreement increases the risk of reaching major climate tipping points by over 70 percent compared to scenarios without exceedances.[16]

The Principle of Hope

What does all this mean? The first conclusion is that a world with 3°C (5.4°F) of warming would have such predictable negative consequences for people and the environment that we can hardly imagine them. If we also consider that the world would go completely out of control due to tipping point cascades, the second conclusion could be: We don't want to imagine such a world. That is what's currently happening. We're collectively ignoring the possibility—and not just those of us who fundamentally consider climate activism to be politically motivated scaremongering. At first glance, this willful disregard may seem surprising. However, the reality is that those who are vigorously fighting for better climate policy often simply avoid even talking about the possible consequences of 3°C (5.4°F) plus scenarios. This includes climate politicians and scientists who work on the United Nations Intergovernmental Panel on Climate Change (IPCC)—i.e., precisely the people who are trying to convince others at the major world climate conferences to advance decarbonization faster and more radically.

There are only a few scientific studies with quantitative forecasts on the possible consequences of a warming of 3°C (5.4°F)—far fewer, in the IPCC reports, than there should be given the high probability of such a scenario. In fact, we're going in the opposite direction: The discussions of the 3°C-plus (5.4°F) horror scenarios are *decreasing*. The view of the institutional climate protectors increasingly concentrates on the desired scenario of 1.5°C (2.7°F) to a maximum of 2°C (3.6°F).[17] This is understandable from a tactical and psychological perspective. No climate scientist or politician wants to be accused of scaremongering, knowing full well that this will make it harder for them to be heard by the many who are needed as allies for climate protection. The target vision of climate policy has been clearly drawn and formulated since the Paris Agreement. A world with 1.5°C (2.7°F) of warming, or "well below 2°C (3.6°F)," is possible if humanity makes a real effort. It is the only desirable world that we still know from today's perspective. That is why the "principle of hope" applies within the consensus-oriented climate policy community. The term comes from the German philosopher Ernst Bloch, coined in the post-apocalyptic mood of World War II.[18] A better world becomes possible if we imagine it because our imagination fuels the willpower needed to actively shape the world toward a desired goal.

I have great sympathy for this attitude. Anyone who, as a politician, scientist, or climate activist, dedicates a large part of their time and energy to keeping climate change to bearable limits must believe in the success of their own work, at least on their good days. How else should they continue? In this context, Bloch also spoke of the "constructive power of concrete utopia." I believe in it, too. Pessimism is a waste of time. It is the precursor to resignation and robs us of the energy to change our environment for the better. But a realistic look at the climate data allows only one other conclusion: The constructive power of the Paris Agreement's concrete utopia is unfortunately too weak. As of today, there is a high probability that our children and grandchildren will have to live in a world with 3°C (5.4°F) of warming by the end of the century. At least that is what current data suggests if looked at without blinders.

The UN Environment Program (UNEP) Emissions Gap Report annually determines the gap between the expected emissions in the coming years and the values necessary to achieve the Paris climate goals. The UN climate scientists presented the last report shortly before the COP28 conference in Dubai. The key indicator is that if all governments in the world fully implement the climate protection measures they have committed to in the Paris Agreement, the Earth will probably warm up by 2.9°C (5.2°F), compared to preindustrial levels of 1850 to 1900, by the end of the century. A 2.5°C (4.5°F) scenario is likely only if the conditional commitments are kept.[19] The trend in the forecast is extra worrying. In 2022, the UNEP assumed a warming of up to 2.8°C (5.04°F). The Climate Action Tracker, an independent consortium of climate scientists, comes to a similar conclusion. It estimates the most likely temperature in 2100 to be 2.7°C (4.86°F) above preindustrial levels if the internationally determined contributions are fully adhered to by thirty-nine countries and gives a statistical range of plus 2.2°C (3.96°F) to plus 3.4°C (6.12°F).[20] How bad will the long-term forecast still become? The days of practical optimism are definitely over. We already live in a world of warming of 1.5°C (2.7°F).

The 1.5°C (2.7°F) Mark Has Already Been Crossed

In 2023, average temperatures worldwide were 1.48°C (2.66°F) above the average of preindustrial levels. Every single day, the average temperature was more than 1°C (1.8°F) higher; on some, it was more than 2°C (3.6°F). From March 2023 to February 2024, the 1.5°C (2.7°F) mark was finally exceeded. Scientists assume 2023 was the warmest year in at least one hundred and twenty thousand years. The most recent surge was caused primarily by the rapidly rising temperatures of the oceans, so-called marine heat waves. In the Gulf of Mexico, near the coast of Florida, temperatures exceeded 38°C (100.4°F) in July. This is doubly worrying because the world's oceans have so far been able to absorb a large part of the heat generated by the greenhouse effect. A heat wave in Western Antarctica caused an iceberg measuring over 1,500 square kilometers (about 580 square miles) to break off from the Brunt Ice Shelf and it now floats in the Weddell Sea. NASA's Jet Propulsion Laboratory researchers recently calculated that, since 1985, the Greenland glaciers have melted significantly more than previously assumed.[21] Forest fires raged in Canada, the US, and large parts of South America in 2023, once again breaking the records of the past few years. The number of climate-related disasters with billions of dollars in damage climbed to a record of twenty-eight in the US, resulting in a total loss of around one hundred billion dollars.[22] The Panama Canal Authority had to radically restrict operations on the waterway due to low water levels. The suffering of the people living in Derna, Libya, cannot be quantified. In September, a "flood of the century," triggered by the Mediterranean Storm Daniel, claimed the lives of more than eleven thousand people.[23]

The world's public reacts to the terrible news with customary dismay. The UN and individual states send disaster aid. On the other hand, world climate policy concluded the climate-horror year of 2023 at the COP in Dubai with a weak declaration of a transition "away from fossil fuels in energy systems in a just, orderly, and equitable manner." At least the delegations from Saudi Arabia and other oil-producing states appeared satisfied after the climate summit, as the declaration "do[es] not affect our exports, do[es] not affect our ability to sell."[24]

A look at the increase in global carbon dioxide emissions from fossil fuels completes the picture. In 2023, emissions were 36.8 billion tons, around 1.1 percent higher than in the previous year and significantly above the level before the outbreak of the pandemic in 2019. The main parties responsible for the increase are China (4 percent of the global total) and India (8 percent).[25] At the end of 2023, the Intergovernmental Panel on Climate Change (IPCC) assumed that total greenhouse gas emissions by 2030 would be 9 percent higher than in 2010, and thus called into question individual studies that predict the peak emissions expected at the end of this decade.[26] Meanwhile, we see more and more countries water down or postpone their savings commitments. Post-Brexit Great Britain has taken a pioneering role in the regression of climate protection and now wants to achieve the transition to the post-fossil era "more pragmatic, proportionate and realistic approach."[27] This includes the fact that cars with combustion engines may now be sold in Great Britain until 2035, not 2030 as initially decided. Requirements such as replacing oil and gas heating systems with heat pumps and housing insulation were also significantly weakened by former Prime Minister Rishi Sunak.[28] In the United States, Donald Trump has made gas and oil great again. The US's leaving the Paris climate agreement (again) will most likely encourage other governments to dial down their climate efforts as well, even if they keep on paying lipservice to a "green energy transition."[29]

There is a growing impression that politicians and companies around the world find it easy to make long-term commitments to cut CO_2 emissions. However, all countries have difficulty meeting short-term goals, except those that are suffering from recession and/or, like Germany, are reducing their industrial production. The CO_2 concentration in the atmosphere reached a new record high of 419 ppm in 2023.

Let's briefly summarize the status quo. Even if the promised climate protection measures are adhered to, the world is heading toward global warming of 2.5°C (4.5°F) to 3°C (5.4°F). We currently live with around 1.5°C (2.7°F) plus—the often-mentioned 1.2°C (2.16°F) result from calculating the average over the last ten years. Greenhouse gas emissions will likely increase in the next decade. There is no evidence that the world will decide to take significantly more radical climate protection measures. There are some indications that the opposite is the case, with particular uncertainty about how the US's climate policy and climate balance will develop whenever a Republican is in power, given the GOP's open hostility to climate science, and what repercussions this will have internationally.

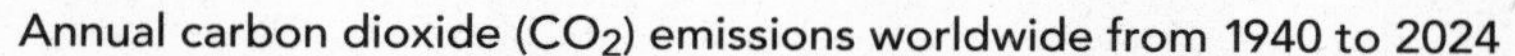
Annual carbon dioxide (CO_2) emissions worldwide from 1940 to 2024

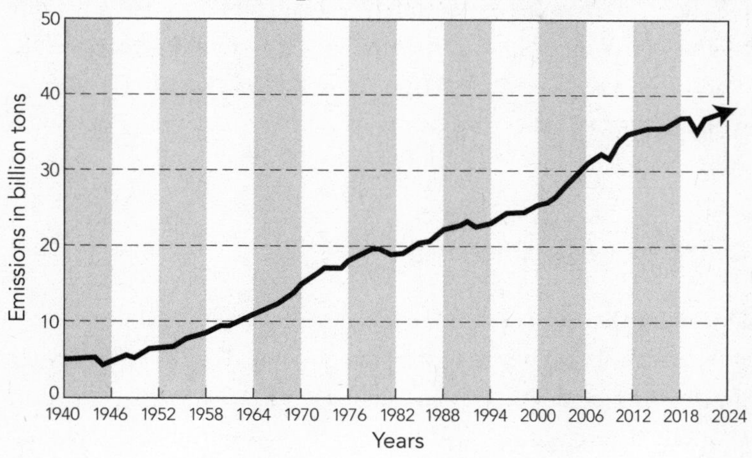

From today's perspective, why should we rely on the fact that the world could be less than 2°C (3.6°F) warmer at the end of the twenty-first century than in the eighteenth century? Optimistic scenarios are helpful. They help us humans to create a desirable future. It is still conceivable that a self-reinforcing climate protection dynamic will begin, which will actually enable warming to be contained to around 1.8°C (3.24°F). Renewable energies are becoming increasingly cheaper, and perhaps nuclear fusion will actually be successful in the 2030s. The major powers may find their way back to a path of cooperation and develop a form of constructive competition: Who will be able to decarbonize faster than the other? Or, to put it more generally, maybe humanity will gain some common sense about climate change and overcome the major "collective action" and "free-rider" problems of climate protection. However, any risk assessment that considers this possible and desirable future to be particularly likely is, in the most generous light, naive. Viewed through a behavioral science lens, one might say it's negligent. In my opinion, it is simply irresponsible. And given the current situation, to not at least seriously grapple with solar geoengineering seems collectively naive, negligent, and irresponsible to me.

Don't Say "Geoengineering!"

Dimming the sun is taboo in the climate policy establishment—not just the action but even the thought of it. The IPCC avoids the topic with such dogmatic consistency it is more reminiscent of the Catholic Church in the Middle Ages than scientific and technical discourse on the options for action in a potentially existence-threatening situation for human civilization.

There are several good reasons to view solar geoengineering with healthy skepticism. Some are geophysical in nature and show that artificially shading the sun's energy input comes with risks and undesirable side effects for the Earth's system that are difficult to calculate. However, the more serious objections are based less on geophysics than behavioral psychology. If solar geoengineering is marketed as a quick solution to climate change, it could become an excuse for irresponsible humanity not to continue the arduous path of decarbonization and to conveniently satisfy humanity's still-growing hunger for energy with cheap coal, oil, and gas. Behavioral economists and social scientists call this the "moral hazard" of solar geoengineering.

I will go into both fundamental problems—in the context of both the natural and social sciences—in detail in chapter 3 and explain why, according to the current state of science, none of the concerns raised can be, in itself, a reason to rule out dimming the sun. Critical questions are now imperative, and when it comes to the possible use of solar geoengineering, so are the typical questions of technology assessment.

Will the use of a new technology do significantly more good than harm?

How would the technology have to be used to benefit many people and, ideally, harm no one?

What incalculable risks does the use of a new technology entail?

People always make decisions under conditions of uncertainty. If, when choosing between several options for action, we know exactly what will come of the choice and, according to the defined criteria, bring more advantages than disadvantages, then we are not dealing with a decision but a routine. Computers can carry out routines better than humans. Large-scale intervention in the Earth system is a huge decision. Before politicians can make it, scientists and technologists have to reduce the uncertainties about the benefits and risks as much as possible and develop the technical use efficiently and safely. Creating the basis for an informed decision in favor of solar geoengineering, or even consciously forgoing it, will take at least ten years, probably longer. The same applies to the question of who could actually make this decision, affecting every person on Earth. And what would be a secure international legal basis to ensure implementation for decades? There are no easy answers to these questions. It's high time to find them.

From today's perspective, there are many indications that solar geoengineering can do significantly more good than harm if temperatures continue to rise. There is little to suggest that, if used responsibly, there is a high risk of chaotic interactions further disrupting the climate, which humans have already heavily "engineered" with its heating. There is much evidence that climate-vulnerable regions can benefit more from solar geoengineering than the more affluent and cooler North. Yet the call for a ban on research remains strikingly popular among climate activists, and research funding remains unavailable. One would think the time of scientific thought control is long gone.

More knowledge and insight are always good. At least, this is the consensus in most sciences. Solar geoengineering is an exception. A part of the climate science community is still skeptical that solar radiation modification (SRM) should even be researched. This undoubtedly also has something to do with the fight over allocating research funding. However, above all, part of the climate science mainstream fears that the price of knowledge about SRM means less climate protection along the known paths to reducing greenhouse gases—and, therefore, the price is too high. This attitude increasingly angers the growing community of geoengineering researchers. Not researching SRM essentially means that humanity is either taking an option to combat climate change off the table, even if the geotechnical intervention could, at some point, represent a significantly lower risk in relation to further global warming, or the decision for or against the use of solar geoengineering would be made on a considerably poorer basis of knowledge than if extensive investment had been made in research beforehand.

In conversations with researchers on solar geoengineering, it is often apparent how tired they are of being pushed into the corner of tech-solutionists who read too much science fiction in their youth. As a rule, they are careful about who they speak to publicly. All of the researchers I talked to for this book preface their statements with the obvious limitations of solar geoengineering and its research: Dimming the sun is *not* a solution, but an interim solution. It buys time for all other climate protection measures. The decision for or against deployment will always be a difficult one. Scientists and technologists have only the task of improving the decision-making basis for those who will make this difficult decision. This caution in rhetoric and argumentation has slowly grown.

When discussing the idea of shading the Earth more broadly and seriously in the years after the turn of the millennium, the campaigning US environmental protection organization ETC Group came across the topic. The group gained global attention with slogans such as "Hands off Mother Earth!" and "Geopiracy." A British experiment that explored the possibility of using a long hose to introduce sulfur particles into the stratosphere was labeled "The Trojan Hose."[30] It was a hit, earning laughter all over the world and becoming a template in content and tone for the world's most global critic of globalization, Naomi Klein. Her blanket criticism boiled down to this message: Geoengineering is a delusion and an attempt by those in power to distract from the actual necessary climate transformation.[31] This is an objection that has repeatedly popped up for several years in the context of postcolonialism. The theme: A technocratic elite in the rich North will claim the decision on the design and use of solar radiation modification for itself and will always prioritize its own interests over those of the Global South. These remain the critics' core messages to this day. It is often pointed out that Bill Gates is "investing" in the technology. The nugget of truth in this is that Gates cofinanced a research project at Harvard University. However, he had no influence over the content.

Skepticism of technology is always an important component of technology assessment. Nobody can rule out in advance that the Global North could organize a major geoengineering push at the expense of the Global South. Accordingly, part of the responsible exploration of radical innovations is creating the greatest possible transparency in scientific research, technical development, and political decision-making processes. Only then can misuse and manipulation be prevented in the event of potential deployment. During my research and discussions about this book, I repeatedly asked myself why even small research budgets and cautious discussions about solar geoengineering met such vehement resistance. Is it the size of the project? That's probably part of it. However, especially among "veteran" environmental activists, the approach seems to shake up too many of their firmly held beliefs: Local, decentralized solutions are better than global, centralized ones. In industrial production, it is always better if no waste or pollutants are produced in the first place rather than having to treat or dispose of them later. Therefore, social solutions to problems are fundamentally superior to technical solutions. Solar geoengineering violates all these principles. Accordingly, it sets off triggers that result in the desire not to have to think about an idea at all. It is easier to dismiss it as absurd and put it in the box of human technological hubris, in which the cause of climate change itself lies.

Risk vs. Risk

The prohibition on even considering solar geoengineering as an option is based on a grave error in thinking. It ignores the fact that the possible deployment of solar geoengineering must be considered in a risk-risk scenario—where refraining from its use *also* risks negative outcomes—not in comparison to a scenario in which current climate policy paths lead to success.

Admittedly, the negotiators of the UN climate protection process find themselves in a contradiction that is difficult to resolve. On the one hand, they must agree on goals to which countries whose current and future prosperity is disproportionately dependent on fossil fuels also commit themselves. On the other, the geophysical condition of the Earth requires exceptionally ambitious goals. The negotiators in Paris were, of course, aware of this contradiction. They hoped that a self-reinforcing process, a new green momentum, could begin based on an agreement that was at once exceptionally ambitious and wholly inadequate. Thanks to falling prices for green energy, one country after another will be able to make ever-higher commitments to CO_2 savings, or so the logic goes. International cooperation—at least that's what it looked like to the negotiators of the Paris Agreement in 2015—could reach a whole new level. Ideally, a kind of outbidding competition would begin in which governments can make a political name for themselves with ever more timely net-zero targets for their countries. This optimistic scenario did not come to pass. If we needed any final proof of this, the 2023 climate summit in Dubai provided it. Overall, when it comes to climate, humanity focuses more on the short-term benefit than the long-term burden. Decarbonization is expensive today. The investment will then pay off many times over in a few decades. With solar geoengineering, it is precisely the opposite.

By spending a few billion dollars annually on sulfur in the stratosphere, human suffering and the financial cost of climate change can be drastically reduced in the short term. If an international institution began dimming the sun on behalf of the vast majority of the international community, as massive volcanic eruptions often do, temperatures would stop rising within a few months. Unfortunately, this doesn't help in the long term, just as a cheap painkiller helps a patient only temporarily until the cause of the illness has been successfully combated. Nevertheless, no doctor would ever refuse pain therapy to the patient if the side effects are less harmful than the benefit of reduced pain.

Once again, an interim solution is not a solution. It offers a way to gain time to find and implement a fundamental solution. The debate about solar geoengineering must be conducted from this perspective. Thinking in terms of short-term success is a synonym for irresponsibility and stupidity in today's climate policy debate. The short-term thinking of older people has brought the youth onto the streets Friday after Friday to fight for their future in a stable climate. The short-term thinking of political actors worldwide triggers the forms of protest of radical climate activists. One may or may not sympathize with protesters blocking roads by glueing themselves to the asphalt, but their insistence on long-term climate solutions is understandable as a response to short-sighted, short-term thinking. However, I'd argue that politicians and protestors alike are overlooking the middle ground—the stopgap solution of solar geoengineering, which might serve as a bridge to a long-term solution. Rigidly sticking to a long-term goal might prevent you from being flexible enough to adopt the interim tactics that could help reach that very goal. But before solar geoengineering can be considered a viable stopgap step, we must ensure that it would be used responsibly. And that is by no means guaranteed under the current conditions.

Big Decision, Big Responsibility

Responsibility usually has many dimensions. In the case of solar geoengineering, there are at least four:

Deployment may be considered only if the scientific evidence about the scope, manner, benefits, and risks of use is sufficient to make a well-informed decision. The number of studies is already significantly better than geoengineering skeptics often claim. This is also remarkable because, so far, only very little money has flowed directly into solar radiation modification research. The sooner large research programs are launched worldwide, especially in countries in the Global South, the sooner an appropriate foundation for decision-making could emerge.

Second, a responsible climate protection policy involving solar geoengineering must do everything in its power to, at the least, not slow down decarbonization but, ideally, even accelerate it. Acceleration could be possible if the necessary use of solar geoengineering makes it clear to every political and economic decision-maker how dramatic the situation in the greenhouse Earth actually is. Any political decision-making process in favor of deployment would have to keep the urgency of the other pillars of climate protection at the center of communication. This would include firm reference to all scientific findings that solar geoengineering alongside uncontrolled emissions of climate gases will dramatically worsen the situation for our children and children's children. In a worst-case scenario, solar geoengineering might be used as a quick and poorly prepared emergency measure to prevent a climate element from reaching a tipping point at the last minute.

The third dimension of responsibility aims at a stable political and legal framework. At first glance, this seems unlikely given the current geopolitical tensions, including a new Cold War between the US, China, and Russia, and numerous acute trouble spots. With closer inspection, however, there is reasonable hope that a geoengineering agreement and regime could be the exception to this. In recent years, one of the few policy areas at the UN level that has remained open to political understanding, despite growing tensions, is environmental and climate policy, and progress has been made, albeit too slowly and at too low a level. One of the few interests that all people and their governments worldwide share is that the climate does not go completely out of control, and with it, the world. In an optimistic scenario, the global community could become more cooperative again through a worldwide decision-making process on solar geoengineering.

Fourth, if ultimately, a consensus on climate action favors solar geoengineering, the technical implementation must be carried out cautiously. The approaches to reducing solar energy discussed in detail in chapter 2 all have a major advantage: They are reversible. Their effect can be terminated relatively easily. A responsible increase in radiation reduction would have to be accompanied by strict observation of the desired and possible undesirable effects. A comprehensive monitoring infrastructure would have to be set up in advance, with sensors on the ground, in the air, and in space. If the data indicates that the harm exceeds the benefit, discontinuation would be the responsible decision. In this case, the sooner the better. The longer and more strongly solar engineering cools the Earth, the more difficult it will be to stop.

Considering these four dimensions, a scenario could look like this: Comprehensive research programs will start immediately to largely eliminate the geochemical and geophysical uncertainties of all available solar geoengineering methods and to identify the safest, most effective method. By the mid-2030s, there would be a scientifically sound basis for political decisions. The United Nations would have defined a decision-making mechanism to reach an international agreement on solar geoengineering. To do this, it will have to create new implementing and supervisory authorities similar to the International Atomic Energy Agency (IAEA). It is highly likely that, at this point, the world will be on a clear path to warming of well over 2°C (3.6°F). However, most countries will likely find the goal of net-zero emissions more feasible thanks to further falling costs of renewable energy and better energy storage. The UN could entrust the implementation to an international consortium of representatives from all continents. From 2040 onward, temperatures can be reduced by a few tenths of a degree in a cautious approach—i.e., the peak of human-caused global warming can be eliminated ("peak shaving"). This not only immediately stabilizes the climate and the prevailing weather patterns but, above all, prevents climate elements from tipping. From the 2050s or 2060s, the processes for extracting climate gases from the atmosphere must be scaled up to such an extent that the carbon dioxide concentration is at least reduced to today's level. This can be achieved with abundant green energy, but it will probably take several decades. The Earth's shadowing—most likely with sulfur in the sky—could slowly be reduced again at the end of the century.

An optimistic scenario includes the hope that after the climatic near-death experience, humanity will be significantly wiser, more forward-looking, and more consensus-oriented. Maybe another miracle will happen. Maybe North America and Europe will go net zero in 2040, followed by the rest of the world in the 2050s. That may be enough to limit the temperature increase to well below 2°C (3.6°F), and it may be enough for humanity and the biosphere to adapt to the slightly warmer world. However, I'm not the only one who doesn't believe in that. Most IPCC scientists have also lost this hope. In an anonymous survey conducted by the American journal *Nature* in 2021, around 60 percent of scientists stated they expected the temperature to rise by 3°C (5.4°F) by the end of the century. At that time, only less than 5 percent still believed the 1.5°C (2.7°F) mark could be maintained.[32]

In the foreseeable future, the question will no longer be whether solar geoengineering should be used. It will be *who* does it, *when*, and *how*.

This alternative is so-called rogue geoengineering. Maybe the US, India, or China, a coalition of states in the Global South, or perhaps a tech billionaire could single-handedly dim the sun. Technically, it is easy. This would be extremely dangerous, politically and socially. "Rogue geoengineering" will not become more likely due to research and debate, nor ignoring it and making it taboo. Therefore, facing solar geoengineering head-on, collectively investigating it and better understanding its risks and possibilities, would be a wiser strategy against rogue use of such a powerful technology.

2

TECHNOLOGY:
The Geoengineer's Toolbox

Volcanic Winter

At the end of July 1990, villagers at the foot of Mount Pinatubo alerted the authorities in Manila. Steam and foul-smelling gases escaped from the volcano's slopes—then over 1,700 meters (5,577 feet) high—on the Philippine main island of Luzon. That was unusual. The volcano was considered extinct. The last time it erupted was around the year 1500, approximately seventy years before Spanish conquistadors conquered what is now the Philippines. The geologists and seismologists from the national authority Philippine Institute of Volcanology and Seismology, or PHIVOLCS for short, set off immediately because the villagers were rightly worried. Two weeks earlier, a severe earthquake had struck the region north of Mount Pinatubo, killing over 1,500 people. The obvious suspicion was that the Earth's tremor had also thrown the interior of the volcano into disarray. The scientists' journey was short. Pinatubo is located about 90 kilometers (56 miles) northwest of Manila. After comprehensive investigations, they initially gave the all-clear. The steam and gases appeared to escape from surface reservoirs exposed by small landslides. No threatening changes could be seen beneath the mountain or in the magma chamber with its 100 cubic kilometers (24 cubic miles) of molten rock.

Wherever they do their work, geologists and volcanologists are trained to act cautiously and avoid premature warnings of eruptions. If they often sound the alarm and nothing happens, the affected residents will eventually no longer follow the evacuation calls. In August 1990, the PHIVOLCS experts traveled back to Manila, some probably with mixed feelings. Beforehand, the locals were asked to pay particular attention to the volcano. Six months later, the Earth actually trembled, this time on several occasions, at intervals of just a few weeks. In April, a 1.5-kilometer-long (0.95 mile) fissure opened near the summit. Filipino volcanologists brought their American colleagues from the United States Geological Survey with them this time. Together, the scientists concluded that the risk of a major outbreak was increasing daily. Seismic activity close to the rock surface became stronger. Underground fissures paved the way for lower layers of hot magma to rise. In May, sulfur dioxide emissions increased tenfold to five thousand tons per day, then decreased abruptly at the end of May. This indicated a blockage that was dramatically increasing the pressure in the lava chamber, making a large eruption more likely. At the beginning of June, a so-called lava dome appeared at the summit, a column-shaped elevation made of particularly viscous lava. This meant the highest alert level was called.

At least Pinatubo was clear in its signals. It also gave the authorities more time to evacuate than some local scientists feared. From April to June, sixty thousand people left a 30-kilometer (18.64 mile) zone around the mountain, many of them somewhat hesitantly and reluctantly. It is still unclear how many people did not follow the call. The first major eruptions occurred at the summit between June 12 and 13. The mountain spewed its lava, lava fragments, rocks, and magmatic gases 24 kilometers (14.9 miles) into the air. Several waves of eruptions followed. The ash took the shape of a mushroom cloud. Then, Pinatubo initially seemed to calm down again before erupting with full force on June 15 at 1:42 PM. This time, the eruption column detonated into the stratosphere to an altitude of 34 kilometers (21.12 miles). In some valleys, the lava flow gushed to 16 kilometers (9.94 miles) from the crater. Geologists estimate that the mountain spewed a total of around 10 cubic kilometers (2.4 cubic miles) of lava, rock, and ash. The sky over Luzon was completely darkened over 125,000 square kilometers (roughly 50,000 square miles) for around 36 hours.[1]

To make matters worse, a typhoon raged that same day. Low-floating ash mixed with the water, creating a mud-like slurry that buried villages. Many houses collapsed under the weight. In the end, at least 875 people died in the disaster area, many of whom were crushed by roofs or swept away by the ash mudslides. The evacuation saved the lives of tens of thousands of people. The precise scientific observation of the earthquake and the correct predictions will go down in the history of volcanology as a great success. But the region is still paying a high price. The infrastructure has been largely destroyed, and the fertile fields at the foot of the mountain can no longer be cultivated. Reconstruction took years. However, the effects of this volcanic eruption—the second-largest of the twentieth century—were not limited to the region or even Southeast Asia. In the following year, the temperature in the Northern Hemisphere fell by 0.5 to 0.6°C (0.9 to 1.08°F), and worldwide, by an average of around 0.4°C (0.72°F). The global cooling effect averaged around 0.3°C (0.54°F) in the three-year period after the eruption.[2] From 1994 onward, the effect slowly tapered off.

With the eruption of Mount Pinatubo, as has often been the case throughout Earth's history, our planet practiced its own solar radiation management. The darkening caused by ash clouds after massive volcanic eruptions did probably trigger several ice ages. The volcanic winter that followed the Philippine volcanic eruption was mild by geological standards. However, Pinatubo was the first major eruption after solar geoengineering was conceptualized, and data on geochemical and geophysical phenomena were systematically collected on all continents. Climate models were modernized to a level of maturity that made it possible to calculate, with relative certainty, the cooling effect compared to a scenario without a volcanic eruption. Climate scientists' interest when tackling these following three questions was and is correspondingly high: What happened chemically in the stratosphere after the eruption? How did the Pinatubo eruption affect the global climate—in general and in individual regions? What can we learn from this regarding the possible use of solar geoengineering?

High Fog

The troposphere is the lowest layer of our Earth's atmosphere. At the poles, it reaches a height of around 8 kilometers (5 miles) from the ground and around 18 kilometers (11 miles) at the equator, with slight fluctuations during the seasons. Around 90 percent of the air and almost all water vapor circulate in the troposphere. This is where most meteorological phenomena occur, with its many chaotic processes that make weather and climate so hard to predict and whose connections are not fully understood by science. During small volcanic eruptions, ejected ash and gas mixtures remain here in this "weather layer." They are washed out again by the water cycles within a few weeks, often causing local damage as ash falls. However, they have no long-term effects on the climate. It gets exciting in climate science when the volcanic ejecta finds its way into the layer above, the stratosphere. Things are much more predictable there, and the chemical and physical processes and relationships are correspondingly well recorded. Accordingly, it is possible to reconstruct exactly what happened above the troposphere in the weeks and months after the eruption of Pinatubo.

In a matter of days, the volcano ejected between 14 and 26 megatons of sulfur dioxide (SO_2) into the stratosphere. Although there is much less water in the stratosphere than in the troposphere, there is still enough to initiate an oxidation process. By reacting with tiny water droplets and air, sulfur dioxide forms sulfuric acid (H_2SO_4). Under the conditions of the stratosphere, the sulfuric acid then forms a milky mist of tiny sulfur droplets known in technical jargon as sulfate aerosols. These droplets also sink into the troposphere, but only at a slow pace. First, they remain in the stratosphere for several years, spreading around the globe within a few weeks thanks to the Earth's rotation and stable currents. Every white aerosol is a tiny barrier to incident sunlight and its energy. The droplets reflect a fraction of the light directly back into space.

As a mass of white fog, it deflects most of it laterally and downward as dispersed light, reducing direct solar radiation on the Earth's surface. This is exactly what happened globally with the Pinatubo eruption, although it was advantageous for the distribution of the sulfur aerosol mist that the volcano is close to the equator. A comparable eruption closer to the poles would have had a major impact on the solar energy input only in that respective hemisphere of the Earth.

The idea of using sulfur particles in the stratosphere to reduce the entry of sunlight, thus lowering the average temperature, is not new. The Russian climate scientist Mikhail Budyko thought of it around twenty years before the Pinatubo eruption. At the beginning of the 1970s, he used regionally adapted models to try to predict what climate changes could occur if large rivers in the Arctic were artificially diverted. The albedo effect of the white ice layers (see previous chapter) in the Arctic Circle played a major role here. In this context, Budyko proposed that the Earth could be cooled inexpensively if high-flying aircraft were powered by sulfurated gasoline. However, he did not assume that this could become relevant in the foreseeable future. In climate science, stratospheric sulfur emissions were filed away as an original idea, a thought experiment, under the figurative name: "Budyko's Blanket." However, the "blanket in the sky" was not seriously discussed for at least two reasons. First, science was only slowly coming to terms with the greenhouse effect caused by carbon dioxide and other greenhouse gases, and global warming was still a marginal issue, even among meteorologists, and was hardly perceived as threatening. Second, sulfur dioxide was making numerous headlines as an environmental poison in the troposphere, causing acid rain that increasingly harmed forests, medieval buildings, and human health. To artificially cloud the sun? At a time when politicians in the US and Europe were just beginning to reduce the darkening of the sky caused by industrial air pollution, which had been increasing since the end of World War II? Budyko's crazy idea ran diametrically opposed to the zeitgeist. And for good reason.

Humanity has been practicing solar radiation modification since the beginning of the Industrial Revolution, albeit unconsciously, in deeper layers of the air and with toxic effects on human health and ecosystems. Until the 1980s, exhaust gases from industry, traffic, and heating ensured that up to 5 percent of sunlight did not reach the ground. This effect has been discussed as "global dimming" among meteorologists and climate scientists since the 1950s. After World War II, sulfur aerosols played a significant role in the cooling effect of the dirty sky and were considered to be the cause of cancer and respiratory diseases. The major smog disaster in London in December 1952, "The Great Smog," which caused thousands of deaths, became a symbol of unchecked air pollution and its consequences for people and the environment. The US Clean Air Act of 1963, which was tightened several times in the coming decades, was the regulatory prelude to the positive "global brightening" beginning around 1990, when the sky brightened through the reduction of air pollutants.

However, the subsequent global air pollution control laws, necessary for health and environmental policy, had (and continue to have) an overlooked side effect: They give an additional boost to global warming. Over the last three decades, "global brightening" has increased global warming by 0.3 to 0.4°C (0.54 to 0.72°F).[3] Less smog, particularly in China, India, and emerging economies, will foster this trend. In its 2021 report, the IPCC assumes that the cooling effect caused by "global dimming" in the troposphere currently roughly parallels the contribution of the powerful greenhouse gas methane.[4] The current concentration of methane in the atmosphere contributes about 0.5°C (0.9°F) to the overall greenhouse effect. Climate activists hope that "global brightening" could be balanced by reducing methane emissions at the same rate as cooling air pollution. This is a desirable scenario but, unfortunately, not a plausible one.

Climate protectors widely consider methane reduction as a particularly fast-acting climate protection measure—an ace up the sleeve. However, they largely overlook the "global brightening" effect, which places additional strain on the climate. Of course, it would be nice if rapid reductions in methane emissions slowed down the current trends of global warming. But it makes little sense to then chalk this up as compensation for the brightening effect of less smog, while temperatures continue to rise due to carbon dioxide emissions. Given the unintentionally negative consequences of cleaner air, a sensible course of action would be to move uncontrolled, health-damaging "global dimming" pollutants from air layers near the Earth up into the stratosphere, creating a new, artificial layer of low stratus. First, reflective aerosols have a much stronger effect there, as is known from the cooling effects caused by eruptions such as the one from Pinatubo. Second, the health risk and damage to flora and fauna would also be many times lower: Much smaller amounts of sulfur are required in the stratosphere for a comparable cooling effect, and they sink to the Earth's surface much more slowly and, therefore, in lower concentrations.

The following chapter will discuss the uncertainties and risks associated with stratospheric aerosol injection (SAI) in detail. But first, the question that needs to be clarified is: How could humans trigger and control a mild volcanic winter? The experimental physicist David Keith addressed this question shortly after the Pinatubo eruption while studying for his doctorate at the Massachusetts Institute of Technology (MIT). Over the next two decades, Keith brought the idea of solar geoengineering from the realm of obscurity into serious scientific debate. If this path of SAI becoming a serious climate policy option was to be given a name, it would have to be "David Keith," even though, as a young scientist, Keith actually had completely different ambitions.

From Obscure to Serious

David Keith was one of the leading minds in the working group who developed the first atom interferometer at MIT in Cambridge, Massachusetts, in the late 1980s—a breakthrough in atomic physics to observe particles and gravitational waves. Keith's doctoral thesis on the topic, supervised by David E. Pritchard, caused a worldwide stir. A great career in the world of small particles seemed to be mapped out. However, to the surprise of many colleagues at MIT, Keith decided to turn his back on atomic physics. The reason was moral: Nuclear interferometers have a high militaristic value, especially for nuclear-armed submarines and Keith didn't want to be a contributing party. Instead, he turned to climate science. An initial result was an essay in 1992 with a title that, in retrospect, reads like an exposé for his subsequent research career: "A Serious Look at Geoengineering."[5]

Since then, David Keith has held many positions at renowned American and Canadian universities and has been exploring the technical possibilities of controlling climate change. He accomplished important pioneering work in the 1990s with his research into carbon capture and storage (CCS). When it came to the mass removal of carbon dioxide from the atmosphere (CDR), Keith didn't just limit himself to knowledge and publications. With the support of Bill Gates, among others, he founded the company Carbon Engineering Ltd. in British Columbia, which wanted to radically reduce the price of direct air capture (DAC) and also convert the carbon dioxide obtained into biofuels.

However, the scientifically brilliant, eloquent, and controversial geophysicist has impacted solar geoengineering the most. The achievements of individuals in science are often overestimated—knowledge is a team sport. In the case of solar geoengineering, however, without Keith, we would know much less about the possibility of dimming the sun. Maybe it wouldn't even be an option. In the 1990s, he was the one who repeatedly initiated meetings and small conferences on a topic that was, at best, kindly laughed at by the climate science mainstream and occasionally openly ridiculed. At worst, the pioneers of solar geoengineering were even discredited as henchmen of the fossil fuel industry. Change finally came almost fifteen years after Keith's first "serious look at geoengineering" thanks to a scientific authority that no one laughed at or suspected of lobbying for climate-hostile big business.

In 2006, the Dutch Nobel Prize winner in chemistry, Paul Crutzen, helped solar radiation modification achieve a scientific breakthrough with his much-cited article "Albedo Enhancement by Stratospheric Sulfur Injections: A Contribution to Resolve a Policy Dilemma?"[6] If Crutzen, known for his warnings on the dangers of destroying the ozone layer, and of the consequences of a nuclear winter from nuclear war, had developed it further, it would have proven to be more than just a dangerous pipe dream of misguided technology enthusiasts. In his text, Crutzen in no way advocated solar geoengineering's immediate use. He called for research into its potential. No more and no less.

Keith took advantage of the tailwind from the support of such a climate science authority. At conferences that were still small after the turn of the millennium, he discussed the risk assessment between the undesirable consequences of geoengineering and the foreseeable climate development without SAI with young colleagues such as the climate scientist Douglas MacMartin and the environmental lawyer Edward Parson. More and more climate modelers responded to his request to incorporate SAI into their models and check the results. In 2013, Keith published a small book simply titled *A Case for Climate Engineering*—to this day, an important reference for everyone who works with solar geoengineering. Keith drew up a plan with several partially parallel phases or steps in which solar reflection, as a responsibly planned and implemented technology, could contribute to solving the climate problem if climate change is not stopped in time through zero emissions and CO_2 removal.

Knowns and Unknown Unknowns

The more reliable the predictions, the better the basis for decision-making. This applies to doctors and their treatments, investors on the stock market, and, of course, climate policy. Predictions are difficult in complex systems where many causal relationships are insufficiently understood, processes are unpredictable, and the general conditions are constantly changing. The climate is a complex system. Therefore, predictions are subject to a high degree of uncertainty. The good news is that thanks to five decades of intensive climate research and many billions of dollars in funding, a better understanding of the geophysical connections in the atmosphere, a multiplication of climate data, rapidly growing computing capacities, better forecast algorithms, and the use of artificial intelligence, the climate models are demonstrably better. This applies particularly to global climate models and the predictive power of different scenarios of further global warming due to continuously emitted greenhouse gases. In recent years, the number of model calculations on the desired effects and possible undesirable side effects of solar geoengineering has increased steadily.

Regarding the desired cooling effect, climate scientists are now largely in agreement, based on measurement data from volcanic eruptions and modeling from common forecasting methods: Solar radiation modification *will* work. With ten megatons of sulfur (i.e., ten billion kilograms, or about twenty-two billion pounds) in the stratosphere, the planet could be cooled by around 1°C (1.8°F). As with volcanic eruptions, the artificial lowering would have the most substantial effect in the first year and disappear again within a maximum of three years. The annual dose would probably have to be increased significantly to maintain solar reflection for decades, because sulfur aerosols have the physical property of clumping together before finally sinking back into the troposphere. Larger aerosol clumps, in turn, have a lower reflective effect.

However, a high degree of certainty about the fundamental effect does not mean that SRM scientists today know enough about which would be the best, safest, and most efficient method to mitigate climate impacts. This begins with the question: Where exactly should the sulfur be introduced into the stratosphere? It is conceivable that sulfur will uniformly nebulize the stratosphere worldwide. To do this, the reflective particles would have to be put into circulation at the level of the equator, since the fog would spread quickly thanks to the rotation of the Earth and would have an effect there within a few weeks. After a few months, the aerosols would migrate from there to the poles—thanks to a constant flow of air called the Brewer-Dobson circulation, which rises from the hot equator into the stratosphere, moves toward the cold poles, and descends again. Introducing the aerosols into the tropics has obvious advantages and disadvantages: Since the stratosphere is significantly higher at the equator than over the poles, the particles would have to be released above 17 to 18 kilometers (10.5 to 11.18 miles). This would mean that they stay in the atmosphere longer; an advantage. Nevertheless, the spreading would be significantly more expensive, depending on the method, than at the poles at around 10 kilometers (6.2 miles). However, spreading Budyko's sun blanket evenly around the world would also mean that the cooling effect would have different regional effects. At the equator, the sun shines more often and more intensely because it comes directly from above, year-round. On the other hand, the sun hits the poles at a lateral incidence angle and only every six months in the Arctic and Antarctic summers. Is that an advantage or a disadvantage? It is possible that an even distribution would have a stronger effect on regional weather phenomena and precipitation than the targeted shading of the polar ice caps. However, the latter is an interesting option that geoengineers are discussing, particularly in light of the tipping points.

If sulfur were released in spring in the High North or South, for example, north or south of the 60th parallel, in the lower regions of the stratosphere at an altitude of 11 kilometers (6.83 miles), then the reflection mist would probably remain at the poles. This strategy would aim to cool the polar ice layers and permafrost areas and thus pull the regional emergency brake before the dangerous tipping points at the poles and their global consequences are reached. Both approaches—insertion above the equator and the polar ice caps—seem plausible from today's perspective. At the same time, they are an excellent example of why more research and modeling are necessary—not only with global climate models but especially with granular regional climate and weather models. Only then can the advantages and disadvantages of different approaches to SRM be meaningfully compared and evaluated. The same applies to the question: Does it really have to be sulfur?

When volcanoes shoot sulfur into the stratosphere, the sulfur molecules first initiate an oxidation process. S_2 is a so-called precursor substance. The hazy sulfate aerosols form only after the conversion into sulfuric acid, which raises the following question among geochemists: Is it perhaps more sensible to introduce sulfuric acid directly into the stratosphere? This would mean skipping a four-week oxidation phase in the cooling process, more precise dosing would also be possible, and the aerosols produced directly from sulfuric acid would probably remain smaller and clump together more slowly. This, in turn, increases their reflective power and enables lower dosages. The advantages and disadvantages of injecting sulfur versus sulfuric acid can probably be clarified relatively easily in experiments and model calculations. In any case, this is likely not the most consequential decision, since both paths are possible from today's scientific perspective. It would be more important to clarify the question of whether other precursor substances for aerosol mist might be significantly more suitable.

These include calcium carbonate (calcite), titanium dioxide, and synthetic particles. Diamond dust would probably also be suitable, as it is chemically inert, but without a doubt it would be far too expensive. The search for the perfect reflectors could be worthwhile because, firstly, sulfur and sulfuric acid remain environmental toxins with a residual impact on ecosystems, even if the dosage is significantly lower than today's air pollution. Secondly, sulfur reacts with ozone (see chapter 3). Of course, it would be good if the risk of damage to the ozone layer could be reduced through geoengineering. At first glance, calcium carbonate seems to be the better choice. It is cheap and has no toxic properties. But sulfur has the great advantage that, thanks to the natural "large-scale experiments" carried out by volcanoes, its effects and side effects are well understood by science and can, therefore, be largely calculated. With any other substance, the risk of encountering unknown risks—the famous "unknown unknowns"—increases. In the laboratory—when chemical processes are simulated on a small scale—not even the "known unknowns" can be clarified with sufficient certainty. This requires experiments in the atmosphere, which are extremely difficult to obtain approval for, if at all, even if the most renowned research institutions apply for them. David Keith knows a thing or two about it due to his "Stratospheric Controlled Perturbation Experiment" (SCoPEx).

SCoPEx is a Harvard University research project initiated by David Keith and German chemist Frank Keutsch in 2015 to provide a better data basis to search for the most suitable precursor substance for aerosol mist.[7] Experiments with sulfur dioxide were repeatedly postponed from 2018 onward, mainly due to legal ambiguities. The COVID-19 crisis then put a stop to the planned tests in 2020. In the summer of 2021, Keith and his colleagues wanted to conduct an outdoor test with a few kilograms of calcium carbonate. Since lime has no toxic effects, they hoped to avoid potential controversy. After a long search, the scientists identified the north of Sweden as a suitable location—more precisely, the sky above the Esrange space center in Swedish Lapland. The space company Swedish Space Corporation has released weather balloons into the stratosphere since 1974 and offered itself as a technically competent cooperation partner for the physical distribution. In the first step, only the hardware was to be tested at an altitude of around 20 kilometers (12.42 miles)—a helium balloon and a specifically constructed distribution container with an integrated opening mechanism. The actual test with lime powder was to follow in the fall or, at the latest, the spring of 2022. According to calculations, it would have created an aerosol cloud 1 to 2 kilometers (0.62 to 1.24 miles) long and around 100 meters (328 feet) in diameter. The researchers wanted to closely observe the distribution and reflection effect, so that the new data could determine whether lime had the potential to be an alternative to sulfur. But then, the US researchers felt headwinds from a surprising direction.

The Council of Indigenous Saami in Lapland launched a petition against SCoPEx, supported by media, including Greenpeace and the German Heinrich Böll Foundation. Indigenous associations in the US expressed solidarity with the Swedish Saami because the experiment may not only have direct negative ecological consequences but, above all, it "would violate the sacred relationship between Mother Earth and Father Sky."[8] Radicalization was also well-established in this protest, as in the end, the SCoPEx employees received death threats.[9] The protest was successful. The SCoPEx Scientific Advisory Board recommended postponing the experiment. Keith and Keutsch followed the advice. Climate activist Greta Thunberg congratulated the Saami. To date, SCoPEx has essentially created social scientific evidence on the campaign and outrage mechanisms resisting geoengineering research, but unfortunately, no hard scientific evidence. In the spring of 2024, the Harvard research project was finally discontinued.

How Does Sulfur Get into the Sky?

For small quantities of aerosol particles, such as those in the SCoPEx project, stratospheric balloons filled with helium are the popular choice as a means of transport. The technology is well-tested and financially favorable. Every day, hundreds of balloons with meteorological measuring devices on hooks rise into the upper layers of the air. Some of them travel through the stratosphere controlled by propellers with high-resolution spy cameras. The altitude record for a helium balloon is 53.7 kilometers (33.36 miles), going beyond the stratosphere. In 2012, a nearly 200-meter-high (656 feet) balloon, measuring 55 stories, transported Red Bull–sponsored skydiver Felix Baumgartner safely to an altitude of 39 kilometers (24 miles) in a pressure capsule weighing around 1.5 tons, from where he then plunged to a record depth. A small weather balloon for hobbyists that can reach an altitude of 18 kilometers (11.18 miles) costs less than fifty dollars online. The activist geoengineers from the Californian start-up Make Sunsets also use commercially available weather balloons. Balloons are also being discussed to transport large quantities of aerosol particles, but they have obvious disadvantages. They are difficult to control in high winds or only with great effort. Takeoff and flight paths are heavily dependent on the weather. The location and timing of their landing are hard to predict, and when operators pop them, as is often the case with smaller weather balloons, the remains frequently end up in the ocean as plastic waste and are swallowed by fish and marine mammals. Generally, a few tons of load may not be a real problem for a giant helium balloon. If necessary, the aerosol particles could be mixed directly into the helium gas in the shell and then released by opening the balloon. But it would take a large fleet to deliver ten megatons of sulfur in this way.

An idea that sounds like it was thought up by a group of technically imaginative elementary school students could be cheaper to implement and operate: A huge, stable balloon is fixed to a specific location in the stratosphere using several long ropes. The balloon connects to the ground station through a thick hose. Sulfuric acid is then pumped into the stratosphere at high pressure. As described in the last chapter, British attempts at this technology offered a target for humorous remarks under the slogan "Trojan Hose."[10] But the technical counterarguments could ultimately be more important than any rhetorical mockery, no matter how witty. Because as simple as the structure of the idea sounds, the actual implementation is just as complicated. A study by Aurora Flight Sciences—a subsidiary of Boeing that specializes in highly innovative aviation projects—concludes that under the harsh conditions in the stratosphere, time after time, the devil is in the detail, and the development risk is correspondingly high.[11] The same applies to the adaptation of standard artillery cannons, with which the precursors of aerosols could possibly be shot into the stratosphere, just as volcanoes catapult sulfur into the sky. This concept continues to circulate on the fringes of the geoengineering community. From today's perspective, it doesn't seem promising. Transport with smaller rockets would also be conceivable and technically feasible, but only at high monetary and ecological costs. At the scientific center of the debate is the means of transport Mikhail Budyko already had in mind for his sulfur blanket: airplanes. The technical requirements here are well-known, partly because there are already military aircraft that can fly in the stratosphere. Fighter jets can fly up to 22 kilometers (13.67 miles) in altitude. Spy planes like the Lockheed U-2, which became famous during the Cuban Missile Crisis, are even designed explicitly for stratospheric flights.

Due to their construction, fighter jets do not have a large load capacity. It would, in turn, be technically conceivable to upgrade classic passenger aircraft such as the Boeing 747 with powerful military engines and thus enable them to fly at the necessary altitudes. However, the safest way would probably be to develop specific aircraft based on the design of spy bombers, which also have high loading capacities. According to estimates, the cost for this is surprisingly low. Wake Smith, a former Boeing manager, specializes in this question. Smith has calculated in detailed scenarios that deploying aerosol precursors with a few dozen special tanker aircraft at an altitude of 20 kilometers (12.42 miles) is around 18 billion dollars (according to the purchasing power of the 2020 dollar)—including development and operating costs and with an estimated operational life of the aircraft of several decades. The costs of large-scale engineering projects almost always increase over the years, and likely, this would also apply to geoengineering aircraft and their operations. But even if they double or triple, they would still be extremely cheap compared to all other climate protection and adaptation measures. The costs of damage caused by climate change would exceed the costs of solar radiation modification by a factor of one hundred rather than ten.

The technically simplest and probably cheapest way to transport sulfur into the stratosphere using aircraft would be converted business jets from manufacturers such as Gulfstream, Bombardier, or Dassault, which can fly up to 16 kilometers (9.94 miles) with a decent load capacity. Keith and Wake outlined this option in an article in the *MIT Technology Review* in the spring of 2024.[12] The text reads almost like instructions for smaller countries or private actors who want to start SRM as soon as possible. For example, a used Gulfstream G650 aircraft costs just twenty-five million dollars. At the 35th parallel north and south, the stratosphere begins at 12 kilometers (7.45 miles) above the Earth's surface. In the north, for example, this would be in the regions of Algeria, Iran, or the south of the US. Here, a small fleet of fifteen converted business jets could release around one hundred kilotons of sulfur or another precursor substance into the stratosphere at an altitude of 15 kilometers (9.32 miles) every year. This amount would not be enough to completely stop global warming. However, according to Keith and Smith's calculations, it could at least reduce the current annual warming by a third. The number of fifteen aircraft, as the two authors explicitly pointed out, was chosen arbitrarily. It could easily be tripled.

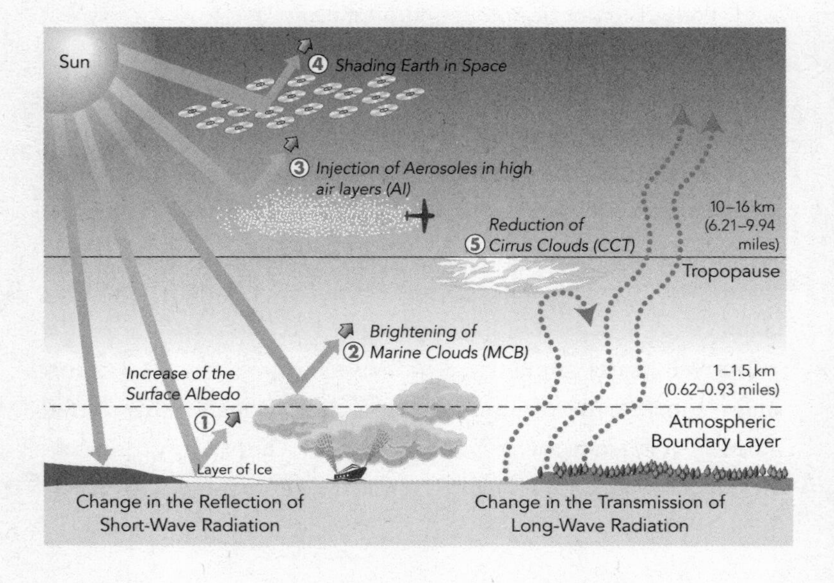

Brightening Marine Clouds, Thinning Cirrus Clouds, and Sunshades in Space

Judging from today's perspective, stratospheric aerosol injection is probably the most efficient method for solar geoengineering. It is undoubtedly the most discussed one. With increasing interest in the research field as a whole, three additional approaches to solar radiation modification have gained attention in recent years. Two of them start below, the other very far above, the stratosphere.

Marine Cloud Brightening

We usually see so-called stratocumuli when we look at the sky on a cloudy day. These are the white clouds below 1,500 meters (about 5,000 feet) altitude that cover the sky in a relatively stable manner. Their lower edges appear gray to us because they absorb particularly high amounts of light. Generally, 20 percent of the world's oceans are covered with stratocumulus clouds. Their cooling effect on the water temperature and, indirectly, the atmosphere is correspondingly large. In the early 1990s, British physicist John Latham, best known for his work on atmospheric electricity and thunderstorms, first suggested artificially seeding marine clouds and enlarging and brightening existing ones. Latham hoped Marine Cloud Brightening (MCB) could be a less invasive alternative to artificial volcanoes.

The basic principle of marine cloud brightening is similar to SAI: Tiny particles are sprayed into the deep cloud layers—not sulfur, but harmless salt crystals. Technically, this is relatively easy, at least on a small scale. Salt water is forced through extremely fine nozzles at high pressure to form a mist. The tiny salt particles (around 100 nm in diameter) serve as nuclei for many miniscule water droplets, making the clouds significantly brighter and increasing their volume to boost their solar albedo.

Several MCB experiments have been carried out off the coast of California since 2009, including by the University of California in San Diego and the University of Washington in Seattle.[13] The results are clear: Marine cloud brightening works—and without the toxic particles in SAI. However, the technical feasibility and its costs have not yet been sufficiently researched. After the turn of the millennium, the original innovator, Latham, suggested having one thousand uncrewed ships circle the world's oceans permanently, powered by environmentally friendly wind rotors. Spray cannons from each ship would shoot 50 cubic meters (about 1,750 cubic feet) of water into the sky every second. A large-scale experiment on the Great Barrier Reef in Australia with such a cannon—similar to the snow cannons on ski slopes—in March 2020 proved that this is also technically possible and leads to the desired result of brighter and additional clouds. It does not seem necessary to spread the saltwater mist with aircraft, which has also been discussed occasionally but would raise the costs too much. According to an estimate by the US National Academies, these costs are significantly lower than those for SAI. One study puts the figure at five billion dollars annually if solar radiation was reduced by five watts per square meter, which would correspond to a global cooling of around 1.5°C (2.7°F).[14]

MCB critics fear, as with stratospheric aerosol injection, unpredictable climatic effects, especially in terms of precipitation, which I will discuss in more detail in the next chapter. It is fundamentally true that with regional interventions such as marine cloud brightening, fine-tuned climate models are necessary to assess the possible risks. Here, the regional models lag the global ones, and to this extent, the effects of MCB have so far been more difficult to predict. However, it is already scientifically certain that to achieve a cooling effect through cloud brightening that is comparable to a stratospheric fog, more regional intervention would have to be made—namely, where the clouds are located. There will most likely be stronger local cooling effects at the selected site. Particularly noteworthy in connection with marine clouds is that the albedo has fallen sharply in the last four years due to an environmental protection measure.

Since 2020, ships in the world's oceans have faced a reduction from a 3.5 percent sulfur fuel cap to 0.5. This has led to an easy-to-observe "brightening effect," as described previously. Since the tankers run on cleaner oil, satellite images of the oceans have shown the fundamentally pleasing effect that the toxic, sulfur-induced stratocumulus clouds, which were visible in satellite images as white bands over the seas, hardly ever form along major shipping routes anymore. The sulfur dioxide from the artificial clouds, a by-product of container ships and tankers, no longer pollutes the air and cannot acidify the oceans after rain. However, the cleaner air drives global warming because the sun is now heating the oceans more.[15]

Cirrus Cloud Thinning

Stratocumuli clouds in the troposphere's lower layers help reduce the Earth's surface temperature. Cirrus clouds, on the other hand, contribute—counterintuitively—to global warming. Forming from ice crystals above 6,000 meters (about 20,000 feet) from the ground, they look like white, transparent streaks in the sky. Despite their position just below the stratosphere, cirrus clouds reflect little sunlight back into space. However, the physical properties of the ice crystals block long-wave infrared rays that are reflected from the Earth's surface back toward space. Overall, this means that cirrus clouds act like an additional natural layer in an Earth glasshouse. If geoengineers completely removed this layer, and the infrared energy coming from below could escape unhindered into space, model calculations suggest that this would have a cooling effect of around 1.4°C (2.52°F).[16] The proposed method for this is technically called Cirrus Cloud Thinning (CCT).

The technical implementation is again similar to that of SAI in the stratosphere and MCB in the lower troposphere. Particles are released into layers of air to change the reflectivity of the clouds. In the case of CCT, these are so-called Ice Nucleating Particles, INPs for short. These ensure that many small ice crystals in the high clouds combine to form fewer but larger ice particles. The clouds initially thin out, and due to the weight of the large crystals, the cirrus clouds sink down even faster and completely dissolve in the form of rain. An advantage of CCT over SAI and MCB could be that it directly weakens the greenhouse effect, and therefore, fewer unexpected climatic side effects are to be expected. On the other hand, as with Marine Cloud Brightening, the cirrus clouds dissolve only where they are located. This necessarily involves unequal local interventions in a global system.

Like the brightening of ocean clouds, the thinning of cirrus clouds is not nearly as well researched and understood as stratospheric aerosol injection. The chemical candidate favored by some geoengineers for stimulating large, heavy ice crystals is Bismuth(III) iodide, a salt without toxic effects that has also been used for decades for weather manipulation in the lower layers of the air—for example, in agriculture or during major events such as the 2008 Olympic Games in Beijing. Sulfur and nitric acid are also being discussed as cirrus cloud thinners. One advantage over stratospheric aerosols would be that conventional aircraft could spray them due to the lower altitude. There is also a lot to be said for bismuth(III) iodide regarding the required quantities. Initial studies show that one hundred and sixty tons of salt annually is enough to offset the CO_2-related greenhouse effect. In case of doubt, a small fleet of adapted transport machines, or drones, would suffice, perhaps even scheduled flights that are already on intercontinental routes. The costs for this are negligible, at least according to initial estimates. David Mitchell, an atmospheric physicist at the Desert Research Institute at the University of Nevada and trailblazer of the climate-friendly approach to geoengineering, estimates the costs at just six million dollars annually.[17]

Shading Earth in Space

The history of technology is full of examples of how science fiction ideas can become reality within a few decades. Perhaps this also applies to gigantic solar sails that partially shade the Earth from space and thus reduce the Earth's temperatures, possibly even specifically at the polar ice caps. The best place in space for this would probably be the L1 Lagrange point, that lies around 1.5 million kilometers (932,056 miles) from Earth on the line connecting the Earth and the Sun. In space dimensions, that's not too far; it is only about four times the distance of the Moon from Earth. It is easy to fly there in spacecrafts, first demonstrated in 1978 by the ISEE-3. At the L1 point, the Earth's and the Sun's gravitational pulls balance each other out so that physical bodies can be placed stably between our heating planet and its energy source—at least in theory. In practice, this wouldn't be that easy. The following plans have so far acted more like thought experiments.

As early as 1989, the American chip developer James Early suggested placing a glass shield with a diameter of 2,000 kilometers (1,242 miles) at the L1 Lagrange point. A 2006 "plan" from the University of Arizona called for suspending sixteen trillion wafer-thin metal disks with a diameter of 60 centimeters (23.6 inches) at L1. However, the researchers themselves immediately questioned the technical feasibility. The theoretical calculation was as follows: If the disks could be limited to a weight of just over one gram, around twenty million tons would still have to be transported there. Even with the current rapid progress in space travel, this hardly seems feasible or affordable. However, according to a newer concept that involves giant space bubbles made of ultra-thin silicone films, the weight of a parasol in space could drop to one hundred thousand tons. As the load capacity of the latest rockets advances, this might become conceivable at some point. The space bubbles of the Italian MIT scientist Carlo Ratti can be inflated to the size of Brazil, could reduce solar radiation by 1.8 percent, and can be maneuvered according to the need for shading—at least in theory.[18]

Technically, it might be easier to cover the Earth with extraterrestrial dust. In 2012, Scottish scientists proposed guiding dust from asteroids to the Lagrange point. Moon dust might also work, as it is literally closer, and the particles have particularly good reflective properties. Astrophysicist Ben Bromley from the University of Utah is currently exploring this idea. In a study from 2023, he simulated different scenarios with electromagnetic cannons or jets that shoot large amounts of moon dust into Earth's orbit that create a giant cloud at the Lagrange point. In initial calculations, Bromley assumes a need for one hundred million tons of dust annually.[19] As expected, such ideas attract an interesting mix of technological optimists and utopians. In this case, David Chaum, one of the first Bitcoin entrepreneurs, found traction with his AstroCool project, which also relies on moon dust at the Lagrange point. Chaum is already fine-tuning technical concepts based on existing space technologies and is preparing a global online referendum to mobilize support for the possible deployment.[20]

With space-based shadowing, the ozone layer would not be affected, atmospheric currents would probably remain unchanged, sulfur would not fall as acid rain, and the sky would not be permanently cloudy. We on Earth would probably not notice a dust nebula in space, except in the form of lower temperatures. While the advantage of space-based approaches are lower geophysical and geochemical impacts on the Earth system, if solar geoengineering measures are adopted in the coming decades, space-based shadowing of the Earth is the least likely method. The technical challenges of the implementation are still too substantial, and knowledge about risks and side effects is limited.

One caveat remains: All approaches to solar geoengineering involve risks. The decision about if and how, as well as the deployment and method, will remain challenging to weigh up. It is all the more important, therefore, that climate science, politics, and the public deal more intensively with the risks of geoengineering in a scientifically sound and open-ended manner. The good news is that this is exactly what most scientists who are fundamentally considering solar geoengineering are doing. The next chapter reports on the legitimate doubts and possible side effects that must be considered, researched, and minimized for deployment.

This book focuses on solar radiation management, the controversial side to geoengineering. As has been emphasized several times, this can serve only as an interim solution to gain time for other climate protection measures. A critical component of the *actual* solution will be the largely uncontroversial part of geoengineering today: removing carbon dioxide from the atmosphere and its long-term storage. In the following pages, I would like to give a brief overview of the tools used in non-solar geoengineering. Because dimming the sun and removing carbon from the air are always two sides of the same coin.

Carbon Dioxide Removal (CDR)

The Orca carbon capture plant near Reykjavik sucks around four thousand tons of carbon dioxide out of the air every year. The gas is mixed with water and pumped into the basalt soil of Iceland. There, it reacts with minerals and crystallizes into calcium carbonate within two years, therefore bound for the long term. The process is complex and energy-hungry. The current concentration of around 420 ppm of carbon dioxide in the atmosphere is far too high and causes the greenhouse effect. However, 420 particles per million are not many if they are to be filtered out of the air using direct air capture (DAC). Due to the chemically low concentration, a lot of air has to be moved, which costs a lot of energy. In Iceland, this process is climate-neutral, thanks to geothermal energy. Therefore, the Orca system is actually "net negative." However, only thanks to cheap energy, favorable conditions for CO_2 fossilization in the ground, and extremely wealthy and willing customers can the pilot plant of the Swiss start-up Climeworks actually pay off. According to the company, sequestering one ton of CO_2 in basalt costs as much as a thousand dollars. Customers like Microsoft and Swiss Re pay the pioneers a goodwill price of around one thousand dollars per ton. At the facility's inauguration in 2021, the Climeworks founders made the optimistic forecast that they could reduce the costs of their process to less than one hundred dollars through scaling. This would (presumably) bring them into a cost range in which they could compete with other CDR methods.

Until a few years ago, some climate politicians and scientists viewed carbon dioxide removal skeptically, including those at the IPCC. As with solar geoengineering today, the fear that dominated for a long time was that CDR as a technical option would be used as an excuse for no longer having to reduce emissions drastically and quickly. The IPCC now has a broad consensus that a net-zero target without high investments in CO_2 removal and technological innovation is unrealistic.

While carbon dioxide removal is the uncontroversial side of geoengineering, there is still uncertainty (and disagreement) about which CDR approaches should be supported and how and if they have the potential to bind many gigatons of carbon dioxide in the long term.

The main approaches for CO_2 removal are as follows.

1. Afforestation, Reforestation, Peatland Restoration

Planting trees is the easiest and cheapest way to remove CO_2 from the atmosphere. Humans sow plants and manage forests sustainably. Nature does the work thanks to photosynthesis and by storing carbon in wood, at least in the medium term. The sun provides the necessary energy. Of course, this has been happening for millennia and on a reasonably large scale. According to the report "State of Carbon Dioxide Removal,"[21] afforestation, reforestation, and forest management cause around two gigatons of negative emissions every year, the case when more CO_2 is removed than replaced. This also includes biochar, which is created by heating wood and biomass in the absence of air (pyrolysis). For comparison, new technical processes such as DAC currently remove less than 0.01 gigatons of CO_2. For further negative emissions, new forest plantings and adapted forest management are possible and desirable in the short term, but problems arise here, too.

The method is only useful against climate change if the CO_2 remains stored in the biomass for many decades at least. However, forest fires and beetle plagues are increasing due to global warming. Negative emissions can be achieved, for example, by using more wood in building houses or by preserving and storing wood underground. Wooden houses makes economic sense, but storing wood safely is expensive. Further, what is often overlooked is a feedback effect that is directly linked to the radiation management of the Earth system. Enormous reforestation projects with gigaton potential are particularly suitable for steppe landscapes or deserts. When unplanted, the light soil reflects a lot of sunlight and cools down significantly at night. On the other hand, a dark forest hardly reflects anything and thus worsens the Earth's albedo. In the short term, large-scale reforestation would warm the climate more through less reflection than it cooled by reducing the CO_2 concentration. However, the negative albedo effect has little impact when it comes to preserving, restoring, and maintaining peatlands and wetlands. They are a previously underestimated method of binding up to two gigatons of CO_2 naturally and cheaply.[22]

2. Bioenergy with Carbon Capture and Storage

BECCS—the abbreviation for "bioenergy with carbon capture and storage"—combines energy production through plants using CCS technology. The highly concentrated CO_2 can be separated from the smoke from power plants much easier than from the air. In this way, energy from fossil fuels can theoretically become CO_2-neutral while the captured carbon dioxide is safely stored. Storage facilities can be, for example, disused gas fields. The same process becomes "net negative" with biofuels because the burned plants first remove the CO_2 from the atmosphere. According to estimates by the IPCC and the US Academy of Science, around five gigatons need to be removed through the burning and fermentation of biomass to achieve the goals set by the Paris Agreement.[23]

The advantage of the process: It helps to make the transition to a post-fossil energy economy easier because the classic power plant structures can be maintained, and net-negative base load electricity can be generated. This is also a so-called Carbon to Value approach: The generated energy can be sold. This is fundamentally more attractive than just selling compensation certificates, as is standard practice today with direct air capture. The question is: When and with what biomass—and at what storage cost—will BECCS become a profitable business if the operators could additionally sell CO_2 certificates? Critics fear that lucrative business models with BECCS will displace agriculture, particularly in the Global South, and endanger food security. That is conceivable, but farmers worldwide could benefit tremendously from their crops becoming, in part, energy producers, or biomass could be obtained incredibly cheaply from aquatic plants and seaweed.[24]

3. Enhanced Weathering (Ocean Alkalinity Enhancement), Ocean Fertilization, Electrodialysis of Seawater

Over 90 percent of the Earth's crust consists of silicate minerals. When silicates (like basalt) weather, they bind carbon dioxide. This geochemical process is complex and varies depending on the type of rock. In nature, it occurs incredibly slowly. However, silicate weathering can be dramatically accelerated if the stones are ground into rock dust to increase the surface area and then scattered over land or oceans. The stone dust particles then react with water and air on land within months to form solid carbonates or in the oceans to form dissolved bicarbonates.[25] This process has the additional advantage of counteracting ocean acidification. For a long time, the potential of enhanced weathering in climate protection was hardly considered. That's changing, especially among scientists. The process could be economically attractive due to its low costs and could become an exciting alternative to traditional mining, especially in the Global South.

For enhanced weathering to ultimately benefit the climate, the mining of the rock and the production and distribution of rock dust must be energy-efficient and CO_2-neutral. On land, enhanced weathering can negatively affect flora and fauna as the pH value of the soil increases. In the oceans, the process using limestone is particularly quick and efficient and, according to studies, has hardly any adverse ecological effects.[26] In this respect, the approach differs greatly from ocean fertilization, for example, with iron or phosphates, which stimulates photosynthesis in phytoplankton. The plankton then converts the carbon dioxide in the water into carbohydrates, some of which sink to deeper ocean layers and are bound there, at least in the medium term. On the other hand, seawater desalination plants can extract CO_2 from seawater using electrodialysis with relatively little additional energy. The waste product of expensive drinking water then results in cheap negative emissions.[27]

The oceans already play a critical role in CO_2 storage today, in the double sense of the word. The oceans absorb about half of human carbon dioxide emissions. As a result, they slow down the greenhouse effect considerably—but unfortunately, they become acidic. If humanity actually manages to become net negative in the second half of the century, the oceans will release some of the stored carbon dioxide.[28] This is desirable for the maritime ecosystem. However, this rebound effect also means that around twice as much CO_2 has to be removed from the atmosphere than it seems at first glance. In this respect, we are leaving a doubly difficult climate legacy for the next generations, and for CDR innovators.

3

RESEARCH:
The Risks and Side Effects

The Desire for Thought Control

"We call for immediate political action from governments, the United Nations, and other actors to prevent the normalization of solar geoengineering as a climate policy option. Governments and the United Nations must assert effective political control and restrict the development of solar geoengineering technologies at planetary scale. Specifically, we call for an International Non-Use Agreement on Solar Geoengineering."[1] An open letter dated January 17, 2022, entitled: "Solar Engineering Non-Use Agreement," begins with these three sentences. It was written by a group of climate and political scientists led by Frank Biermann, Professor of

Global Sustainability Governance at Utrecht University, and Aarti Gupta, Professor of Global Environmental Governance at Wageningen University, who for many years, has been one of the most prominent critics of geoengineering. The letter's content is based on the authors' policy paper published on the same day in the specialist publication *WIRES Climate Change*.[2] The open letter is designed as a petition and aimed at national governments, the United Nations, and political publics worldwide—a media kit is included on the website.

On the day it was published, sixty scientists from all over the globe had signed the call for a ban. In the summer of 2024, there were over five hundred. Numerous nongovernmental organizations, including Amnesty International, the WWF, and the Heinrich Böll Foundation (associated with the Green party of Germany), support the initiative. The worldwide media response was and is great and well-documented in the media section of the campaign website. Of course, the "Solar Geoengineering Non-Use Agreement" is not an "agreement" to date, even if the name sounds like it. However, the initiative has developed into a global rallying point for resistance to solar shading research. What seems obvious at the moment is that with more open discourse about geoengineering, protests against it are also growing, especially among climate scientists and international policy experts who are particularly active in the climate discourse. What exactly motivates those who assess the dangers of climate change as particularly high to not consider solar geoengineering?

Biermann, Gupta, and their co-authors consider the research and development of SRM to be "dangerous" for three reasons:

First, the possible physical side effects on the climate and Earth system may never be fully understood. The geoengineering critics state: "Impacts will vary across regions, and there are uncertainties about the effects on weather patterns, agriculture, and the provision of basic needs of food and water."

Second, speculative hope about the future availability of solar geoengineering technologies could discourage governments, companies, and societies from doing their best to achieve decarbonization or carbon neutrality quickly. The letter says that this speculative possibility of future solar geoengineering risks is becoming "a powerful argument for industry lobbyists, climate denialists, and some governments to delay decarbonization policies."

Third, the current global governance system is unfit to develop and implement the far-reaching agreements needed "to maintain fair, inclusive, and effective political control over solar geoengineering deployment. The United Nations General Assembly, the United Nations Environment Programme or the United Nations Framework Convention on Climate Change are all incapable of guaranteeing equitable and effective multilateral control over deployment of solar geoengineering technologies at planetary scale. The United Nations Security Council, dominated by only five countries with veto power, lacks the global legitimacy that would be required to effectively regulate solar geoengineering deployment."

The conclusion from this threefold criticism—physical, behavioral economics, and political—is a prohibition of solar geoengineering. The authors of the "Non-Use Agreement" demand the following in a five-point catalog for national governments.

- The commitment to prohibit their national funding agencies from supporting the development of technologies for solar geoengineering, domestically and through international institutions

- The commitment to ban outdoor experiments of solar geoengineering technologies in areas under their jurisdiction
- The commitment not to grant patent rights for technologies for solar geoengineering, including supporting technologies such as for the retrofitting of airplanes for aerosol injections
- The commitment not to deploy technologies for solar geoengineering if developed by third parties
- The commitment to object to future institutionalization of planetary solar geoengineering as a policy option in relevant international institutions, including assessments by the Intergovernmental Panel on Climate Change[3]

Like other critics from the camp of climate scientists, the authors of the "Agreement" also assume that the decarbonization of all economies is still possible and that the Paris climate goals are, therefore, achievable. They do not consider solar geoengineering to be necessary, much less desirable, ethically justifiable, and politically manageable. Interestingly, the authors also point out in a sidenote that the "Non-Use Agreement" they presented does not, strictly speaking, represent a call to prohibit research as such. This is difficult to understand when looking at points one and two of the ban catalog. The initiative made its stance on freedom of research crystal clear in its so-called Briefing Note, which reached the public a year after its open letter. SRM research per se is viewed as highly dangerous because "more research cannot resolve the social and political risks that come with geoengineering, and more research will not reduce the risk of delaying urgently needed transformative policies." Instead, more research into "this speculative technological 'solution' will not prevent powerful organizations or countries from deploying this technology unilaterally without global consent or oversight." The report continues that new research cannot prevent "the global impacts of solar geoengineering from being inequitably distributed. And new research will not help to alleviate the fundamental challenges of governing, in a fair and equitable manner, the potential future deployment of a speculative technology that carries such complex risks and unequal global impacts."[4] Their key term is "normalization." Critics like Biermann and Gupta are terrified that simply weighing the risks of a world with geoengineering against a world with climate change but no planetary intervention will create a so-called slippery slope, meaning that any available technology will automatically be used.

In the history of humanity, thought control is a continuum. The fear of normalizing heretics' positions through the spread and recognition of radically new thought drove powerful institutions. Yet the desire to prevent knowledge has rarely been voiced from within the ranks of science itself. This makes the fundamental criticism of the Agreement's signatories obscure, but it also gains relevance because it is adopted more or less unquestioningly by parts of the climate movement. Political decision-making is, at least in theory, a rational process in which all the likely advantages and disadvantages of a decision are considered and assessed, and the pros and cons are comprehensively discussed. German energy policy went from a decision to depend on Russian gas to the nuclear phase-out, a particularly striking example of irrational decision-making on climate issues. Short-term and long-term advantages and disadvantages are not systematically weighed against each other. Symbols beat arguments. This makes it all the more important to deal with the risks and side effects—and to do so in the necessary scientific depth. So, let's start with the legitimate physical concerns and possible negative consequences for the Earth system.

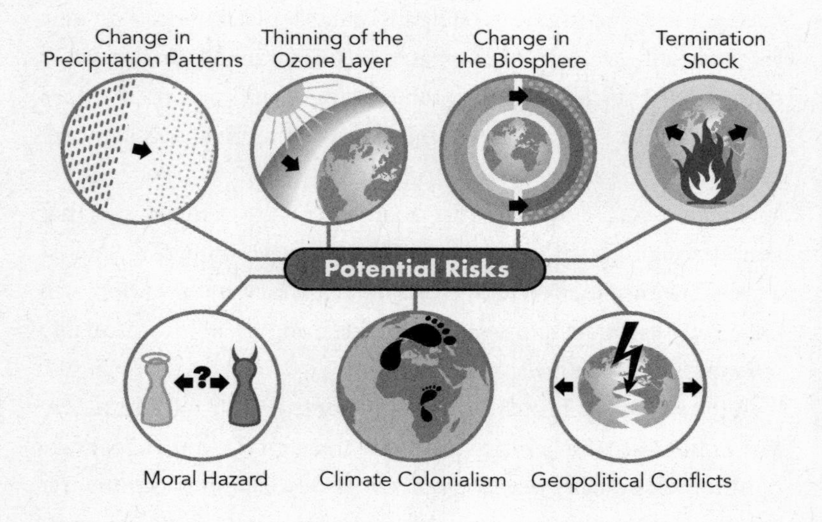

Regional Climate Shifts, Precipitation Patterns, Severe Weather Events

Two things are scientifically proven today. First, solar geoengineering can be used to cool the Earth. Second, solar geoengineering cannot be used to restore the preindustrial climate, to which humanity in the various climate zones has adapted well over the last ten thousand years. The natural greenhouse effect, which is enhanced by carbon dioxide and other greenhouse gases, differs qualitatively from the cooling effect that can be brought about by solar radiation modification. The greenhouse effect, which keeps (now too much) heat in the atmosphere, works day and night. Sun reflection can cool only during the day; this method cannot achieve any effect at night and in winter at the poles. Lowering temperatures is just one aspect of SRM. Particularly through sulfur aerosols in the stratosphere, solar radiation modification will impact the hydrological cycle in the climate system and, as a result, will most likely influence precipitation patterns. The crucial question here is: How much and in what way does precipitation change compared to scenarios with unchecked global warming? Basically, as water cycles change, the risk of severe and extreme weather events increases according to the simple pattern: Where it is wet, it gets wetter. Where it is dry, it gets drier.

Uniformly introducing sulfur aerosols into the stratosphere would certainly reduce the global average temperature. The reduction could most likely be regulated with relative precision by slowly increasing the quantity. However, as is commonly known, the average is rarely a good indicator, especially not on a global scale. An oft-cited comparative study based on the simulation of twelve common climate models concludes that with a uniform distribution of sulfur aerosols worldwide, the tropical climate zones would cool more than the Arctic and Antarctic.[5] The stronger cooling effect near the equator results from the longer average annual duration and vertically incident solar radiation. On the one hand, this would have the advantageous cooling effect that would be particularly effective where people suffer exceptionally from additional heat. On the other, it may be of little use in the so-called cryosphere—i.e., in places where sea and land ice are melting at an accelerated rate and the permafrost areas are in danger of thawing. Today, temperatures at the poles are rising faster than the global average. Therefore, shading the poles, in the first step, could be a smaller intervention of the Earth system, with a relatively higher effect on stabilizing the climate. Ideally, this could help rebuild lost ice sheets, also indicated by model calculations.[6] However, these are, naturally, all just hypotheses for the time being, based on optimized models with increasing prediction probabilities. Critics rightly point to the statistical uncertainty.

The models currently used for climate science forecasts have two inherent flaws. Firstly, our knowledge about the aerosols' reflection effect in the stratosphere is based primarily on observation results from volcanic eruptions. However, in the case of the planned SAI, sulfur particles would not be introduced into the higher air layers by a few powerful thrusts but would be continuously distributed. Secondly, no climate model has yet been developed specifically for predicting solar geoengineering, but standard models have been supplemented and adapted accordingly. Both issues increase the statistical uncertainty. In addition, the natural interactions between the different climate zones must be incorporated into the models. The Earth's atmosphere effectively transports heat over long distances. When temperatures change in one world region, it impacts others. This applies not only to temperatures but also to precipitation, which makes things much more complicated from a climate science and political perspective.

Many skeptics say that solar geoengineering will disrupt African and Asian summer monsoon rains, endangering the food supply of billions of people. But it's likely an invalid point. By no means does it describe the statistically likely scenarios in the case of moderate and planned use, but rather in the event of an exceptionally rapid and strong increase of SRM in the stratosphere until the current warming is entirely compensated for or even corrected to preindustrial temperatures. Nevertheless, the impactful image of absent monsoons and starving children in Africa and India has dominated the debate about geoengineering for more than a decade. The perception has been that anyone who advocates for geoengineering will also cut off the water for poorer nations.

The claim initially came from the US climate scientist Alan Robock.[7] Robock, along with Aarti Gupta and Frank Biermann, is one of the loudest voices against solar geoengineering, with Gupta in particular repeatedly describing in drastic terms the horror scenario of an Asian continent without monsoons. Robock is a volcanologist and an expert on the climate consequences of volcanic eruptions. Maybe that shaped his view of the topic. When the Indonesian volcano Tambora erupted in 1815 with twice the force of Pinatubo, it caused famine in Southern China and the Indian monsoon rains actually stopped the following year.

Most climate models predict that as the climate warms, the Asian monsoons will intensify, often with catastrophic consequences such as those in Bangladesh and Pakistan in recent years, with thousands upon thousands of deaths from floods and mudslides. Most simulations with stratospheric sulfur injection predict that precipitation patterns in Asia will stabilize and that agriculture in Africa and Asia will benefit tremendously because plants will be exposed to less heat stress, especially in the early growth phases. According to calculations by atmospheric scientist Ken Caldeira from Stanford University (initially an SAI skeptic), rice and wheat yields in India could increase by 15 to 25 percent through geoengineering.[8] These studies are, per se, not a sufficient argument for using SAI. If better harvest conditions in India, for example, could be achieved only at the price of less rain in China, the risk of international political disputes would increase. However, the example of monsoons clearly shows that the symbol-laden criticism of solar radiation modification ignores scientific findings—whether consciously or unconsciously. The basic geoscientific and meteorological principles are often ignored, and the potential risks of geoengineering are exaggerated.

One of the key findings of climate science is that global average rainfall is increasing due to global warming. Warmer air can store more water, around 7 percent more for every 1°C (1.8°F) of warming. Warm, humid air allows its water to escape too quickly in the form of heavy rain. Meteorologists are talking about the hydrological cycle strengthening. This is the main reason climate change is already increasing the number and intensity of floods caused by heavy and continuous rain.

Counterintuitively, a boosted hydrological cycle also means that already dry areas receive even less water. The water stored in warm air also rains significantly more heavily over the humid regions. If the air temperature were cooled through solar radiation modification, the increased humidity in the troposphere would be reduced again. Generally, SRM initially offers the chance to stabilize traditional precipitation patterns, a consensus among climate modelers. At the same time, however, the reflecting fog would increase the average temperatures in the upper stratosphere. This, in turn, would definitely impact global heat distribution because, according to assumptions, the stratospheric heat transport would change due to the Brewer-Dobson circulation (the flow of air toward the poles). Based on numerous model calculations, it seems almost certain that a unilateral geotechnical temperature reduction in either the Northern or Southern Hemisphere would shift the so-called Intertropical Convergence Zone (ITCZ). The ITCZ is the broad strip of tropical, rainy climate along the equator. A shift in rainfall toward the warmer hemisphere has been established in the geological history of massive volcanic eruptions. Therefore, we can assume that a serious and well-planned use of SAI will avoid unequal cooling of the North and South. Otherwise, winners and losers would be inevitable, and geopolitical conflicts would undoubtedly arise. If an actor were to actually initiate such cooling unilaterally in one hemisphere, this could at least be compensated for by introducing the same amount of sulfur aerosols into the other half.[9]

In the following chapter, I will discuss the geopolitical landscape of geoengineering—and the options for avoiding conflict. From a geophysical perspective, it can be said that, based on current knowledge, changing precipitation is not an argument against SRM and is by no means a triumphant end to discussion. Instead, the opposite seems to be the case. The prediction models need to be optimized. Meanwhile, the shift in weather patterns in the climate zones caused by unchecked global warming could become one of the most important arguments for SRM, as it could stabilize global water distribution. In this context, in addition to the water storage capacity of the air, evaporation from the ground also plays an important role. Soil moisture is an equally important parameter for ecosystems and agricultural productivity. The major droughts are caused not only by too little rain but, above all, by more evaporation due to higher solar radiation and temperatures. Dimming the sun would mean reducing evaporation, which could benefit fauna and farmers. Further, another factor comes into play here that, on the face of it, seems surprising.

Although sulfur aerosols in the stratosphere reflect some of the sunlight (and its energy) back into space, photosynthesis in plants on the ground would be stimulated, according to current knowledge. As described in the last chapter, the sulfur nebula would divert more diffuse light downward, which, viewed from Earth, might have the disadvantage of giving a cloudless sky a slightly milky haze. For plants, however, more scattered light means more energy to process (and bind) carbon via photosynthesis. This applies to all plants not exposed to direct sunlight, such as those in the lower shady layers of forests and rainforests. The net balance means that this redistribution of light energy will cause more photosynthesis. At least, this conclusion is suggested by comparing plant growth of the twentieth century, from the 1960s to the end of the 1970s, with the era of "global brightening" from the 1980s to the millennium. Because the proportion of diffuse light had increased significantly until the 1950s due to air pollution, plants worldwide absorbed forty-four million tons more carbon annually than after the first air pollution control laws came into force.[10] In this context, however, it should be noted that many crop plants, such as rice, soy, corn, and wheat, prefer direct light. Whether higher humidity with a little more scattered light would be an advantage or disadvantage for agriculture is currently still an open research question. What is certain, however, is that the milky-white sky would reduce the productivity of cells in solar panels by 2 to 5 percent, as they primarily convert direct light into electricity.[11]

Climate Colonialism?

Critics of solar geoengineering research typically assume in their scenarios that its use will deepen the divide between climate change winners and losers. They also believe it is likely that the winners will be in the rich and powerful countries of the Northern Hemisphere because that is where the technology will be developed, mastered, and its deployment controlled—in other words, in their favor and at the expense of climate-vulnerable countries. The group around Biermann and Gupta is also convinced that the Global South will suffer additionally from SRM. Some critics even see the approach as a new form of climate colonialism. Solar geoengineering allows countries "with historical responsibility for the climate crisis to perpetuate their colonial values, power, and politics by controlling access to yet another technology and extending extractivist economics and fossil fuels."[12] This is a possible scenario. But is it a likely one? And does a distinction based on the ingrained perceptions pitting Global North against Global South make sense concerning solar geoengineering? I will discuss this in more detail in chapter 4. However, current model calculations give rise to the hope that, if used responsibly, almost all people in nearly all climate zones can benefit—even those considered particularly "vulnerable" today. The reasons for this are the same as why climate-vulnerable countries and regions suffer disproportionately from warming. The poorer states in the South are more dependent on their own agriculture, generally have weaker institutions, and are less able to invest in climate resilience due to tight national budgets. Conversely, this means that if the temperatures in the Global North and South are moderated equally, the relative advantage for warm and poor regions is relatively higher from a purely economic perspective—which is also desirable from a justice standpoint.

In this context, David Keith refers to his own model calculations in which the change in precipitation with SAI across twenty-two climate zones can be reduced by almost 90 percent—compared to scenarios without SRM, extrapolating the ongoing climate change we have been seeing for decades now.[13] That itself would be a significant benefit. Perhaps even more exciting in this context are the econometric studies by the climate, environmental, and energy economist Anthony Harding and the climate scientist Katherine Ricke. They conclude that solar radiation modification could reduce international income differences, especially those between the rich countries of the North and the poor of the South.[14] Keith and Harding recently presented a study showing the health advantages and disadvantages modeled in a risk-risk comparison of scenarios with and without SRM. The result here is crystal clear: With 2.5°C (4.5°F) of warming, around one million more people worldwide would die every year than if the temperature were lowered by 1°C (1.8°F) using solar geoengineering. People in warm and poor regions would benefit more from this effect than those in cooler and more affluent countries.[15] This is only a model calculation, and the statistical uncertainty exists here, too. However, in contrast to the critics' postcolonial thought patterns, the authors offer a scientifically based design perspective, not just gloomy scenarios without solar geoengineering. And while this is not a sufficient reason in itself to use the technology, the assumption that dimming the sun automatically leads to postcolonial mechanisms that once again favor the North is not a convincing justification for thought control and sanctions on research. Because there is a good chance that everyone can benefit—the Global South even more than the rest of the world.

A New Ozone Hole?

On the ground, especially during the summer when the sun heats up the smog in the cities, high ozone levels harm us. In the stratosphere, however, the ozone layer protects life on Earth from ultraviolet radiation, particularly UV-C and UV-B radiation, which can penetrate organic tissue, damage DNA, and cause cancer. Anyone who lived through the 1980s will remember the discussions about the hole in the ozone layer—and the fear of soon no longer being able to safely go out in the sun. The cause is human. Since the 1930s, we have been releasing large quantities of chlorine and bromine compounds into the atmosphere, mainly in the form of so-called chlorofluorocarbons (CFCs) from refrigerant gases and spray cans. In the stratosphere, these compounds destroy the Earth's natural UV protection shield because their reactive radicals combine with ozone and break it down—a process that has been particularly noticeable over the poles since the 1940s. However, the first warnings from British scientists in the 1950s received little attention. In 1985, a large ozone hole was discovered over Antarctica, and the international community finally woke up to environmental policy. The 1987 Montreal Protocol led to a global phase-out of CFC gases. Since then, the ozone layer has been recovering, albeit very slowly, because the depletion of the ozone killers in the stratosphere is sluggish. Despite the successful CFC ban, the ozone concentration will probably have reached its preindustrial level again only after 2050. This is of particular importance for stratospheric radiation modification. Sulfur aerosols react with chlorine and bromine. Due to these complicated reactions in the gas mixture of the stratosphere, the currently ongoing rebuilding of the ozone layer will, at best, slow down due to SAI, but possibly even open new ozone holes over the poles. The calculation is simple: The earlier sulfur aerosols are used, the more chlorine and bromine compounds are still in the stratosphere, and the harder it is for the ozone layer to rebuild.

After the Pinatubo eruption, the ozone layer thinned by around 4 percent along the tropics and mid-latitudes, which had no measurable impact on life on Earth. The effect was significantly stronger at the poles. Even if SAI were used, the negative effects at the poles would be significantly greater than in the middle latitudes, as chlorine and bromine compound concentrations are still highest in the Arctic and Antarctic. Meanwhile, there is also a counter-effect against the increased UV radiation caused by less ozone; the sulfur particles reduce direct sunlight, especially UV-C, through their reflection. It is difficult to predict how strong these effects would be. However, one thing is clear: The interactions between sulfur, chlorine, bromine, and ozone pose risks that still need to be researched and quantified in more detail.

If the researchers concluded that the ozone layer was at too great a risk, other aerosol precursor substances such as calcite or titanium could be considered, presenting new challenges. According to current knowledge, the brightening of ocean clouds and the thinning of cirrus clouds would not negatively affect the ozone layer. But as a reminder: The debate on solar geoengineering gained momentum in 2006 because Paul Crutzen, the most important expert on ozone depletion, considered the risks of additional damage to the ozone layer caused by sulfur aerosols to be less dramatic than is often claimed. Under no circumstances is it an exclusion criterion for deployment. This also applies to the so-called termination shock.

Termination Shock

The term "termination shock" was not popularized by scientists but by the American science fiction author Neal Stephenson. In his novel of the same name, he describes (similar to Kim Stanley Robinson in *The Ministry for the Future*) a world in disorder by climate change with unbearable heat and major floods. In the novel, there is no geoengineering ministry, so a Texas oil billionaire single-handedly dumps sulfur over the US, which leads to major political complications, among other things, because the monsoons start failing in India. An Indian mission wants to forcibly stop the deployment with a swarm of combat drones, a reaction that leads to a new problem—the very termination shock of the title, a rapid warming to original temperature levels. This (like many other things in Stephenson's novel) has scientific basis.

As already described, sulfur aerosols from the stratosphere sink back into lower layers of air over a time frame of up to three years. This is initially good news because it means the effect is essentially reversible. But if the deployment of sulfur stops, the resulting cooling effect will end, at the latest, three years later. The Earth will then heat up to the temperature it would have reached without geoengineering, depending on the CO_2 concentration in the atmosphere, but now much faster. However, warming that is accelerated many times over is much more difficult for people and ecosystems to endure than slow and steady warming that allows time for adaptation. An abrupt termination of solar geoengineering could have shock-like consequences, causing particularly severe droughts and precipitation and leading to even greater political and biological upheaval. The negative consequences would be correspondingly harsher than in a world in which neither a community of states nor, as in Stephenson's case, a self-righteous Texas oil billionaire had fiddled with the Earth's thermostat.

Skeptics of solar geoengineering assume that SAI would have to be continued indefinitely once started.[16] Their main argument here is that solar geoengineering will almost inevitably bring the decarbonization of the global economy to a standstill. From a geophysical point of view, the danger of "termination shock" is also a consensus among supporters of solar radiation modification. In this context, their answer is again that solar geoengineering is not a solution but rather a bridging technology that buys humanity time for decarbonization. For the deployment, this means that, first, the reliability of the output must be ensured for many decades. Even an interruption of just one year due to technical problems could trigger a "termination shock," which means that the release of aerosols would have to be secured in multiple redundant ways. Second, the cooling effect must be ended slowly and gradually, similar to medications such as cortisone, which cannot be stopped abruptly by patients, but must be tapered off. This tapering of SAI should occur at the same rate and pace as progress is made in reducing the greenhouse effect.[17]

The Best Excuse to Refrain from Climate Protection?

"Geoengineering holds forth the promise of addressing global warming concerns for just a few billion dollars a year." The statement comes from Newt Gingrich, the controversial and contradictory former Republican speaker in the House of Representatives. Politically, Gingrich offers a strange mix of Christian, ultra-conservative, and technologically optimistic perspectives. He was the great moral prosecutor in ex-President Bill Clinton's sex affair yet is an ardent supporter of Donald Trump, whose treatment of women has been illuminated in court several times. Support for the fossil energy and mining industries was a constant in his lengthy political career. In 2008, Gingrich sang the praises of geoengineering as a quick solution, among other things, in a written response to a proposed new climate law. After the big promise of a few billion dollars, he wrote: "We would have an option to address global warming by rewarding scientific innovation. Bring on American ingenuity. Stop the green pig."[18] Newt Gingrich still has influence on Republican research policy. During Donald Trump's first term in office, he lobbied for an extensive research program, but nothing came of it, even though Trump was said to be open to the idea. In his second term, the issue will certainly be put back on the table.

Newt Gingrich is the personification of what climate scientists call the "moral hazard" of solar geoengineering. In behavioral economics, a "moral hazard" describes a lack of incentive to protect oneself from risk because one is supposedly safe from possible harm. For example, the protection can be theft insurance if a bicycle owner buys only a cheap cable lock, not an expensive granite U-lock, or a skier who feels (too) safe with a helmet and back armor and then skis (too) fast. Both are examples of moral hazard. A spokeswoman for the ETC Group sums up the moral hazard in response to Gingrich's lobbying efforts as follows: "In their view, building a big beautiful wall of sulfate in the sky could be a perfect excuse to allow uncontrolled fossil fuel extraction. We need to be focusing on radical emissions cuts, not dangerous and unjust technofixes."[19] Hence, most geoengineering research proponents I spoke to agree that the "moral hazard" is actually a problem that needs to be considered and solved. But here, too, things are more complicated than the skeptics believe or at least say.

The Austro-American climate economist Gernot Wagner from Columbia Business School first points out that the "moral hazard" of geoengineering is, strictly speaking, a "mitigation deterrence"; i.e., it prevents the implementation of expensive and inconvenient climate protection measures, especially decarbonization.[20] Gingrich and, possibly soon, Donald Trump frame the technical option precisely in the sense of hindering a climate policy that they believe is a thorn in one's side. This "either-or" logic triggers applause from the wrong side for the technical option of solar geoengineering, which is grist to the mill of those in favor of a de facto research ban.

Climate and behavioral science show clearly that solar geoengineering is not a technical solution to humanity's biggest ecological problem. Any technofix attitude must be politically combated and contained. This should also be possible in international climate policy because such a belief is obviously wrong. However, the more exciting question in the long term is: Could the discussion about solar geoengineering simply serve as an excuse for those who do not ideologically question decarbonization (like some of the Republican party), but who just want to weaken the current climate protection goals and put off implementing expensive climate protection? Of course, this suspicion is obvious and corresponds in many respects to our everyday experience of human nature.

Most of us often put off unpleasant tasks, even if we know that doing so will make the task more urgent and challenging. As we all know, welcome excuses to postpone unpleasant decisions are also prevalent in politics, observable before practically any major reform that might be met with low approval by voters. Not only activists but also political scientists fear that solar geoengineering follows the logic of either-or.[21] However, recent studies, mainly from game theory and behavioral economics, indicate that the moral risk of geoengineering could also result in a productive "yes, and" logic if the political framing is correct. The central idea is that a serious discussion about solar geoengineering as a last resort measure against climate collapse emphasizes the urgency of decarbonization.

Experiments by social scientist Christine Merk suggest that this logic is more than wishful thinking. In large groups of subjects, she tested whether knowledge of geoengineering reduces the likelihood that a subject will donate part of their fee for participating in the study to climate protection projects. The opposite was the case: When geoengineering information is presented as the last resort of climate policy, it increases the desire for decarbonization.[22] The same could apply not only to a better-informed public but also to those who negotiate climate protection nationally and in international organizations and committees. In the spirit of classic realpolitik, the threat of starting solar geoengineering only on a national level could even become a bargaining chip, as particularly vulnerable countries, such as island states, pressure the latecomers to take stricter climate protection measures. As early as 2010, the economist and urban planner Adam Millard-Ball from UCLA hoped, based on a game theory model, that the "collective action problem" of climate protection could be solved this way and called it the "Tuvalu Syndrome"—a reference to the country that might become the first to disappear from the map due to rising sea levels, and which therefore might take its chances in unilateral geoengineering action.[23] Of course, this has not happened yet.

However, looking back at the past decades, it is revealing that "moral hazard" was repeatedly used as a counterargument against climate protection measures that are now the consensus. In the 1990s, proponents of climate adaptation measures had to fight against the argument that such adaptation would provide climate sinners with excuses free of charge. Adaptation strategies are now a natural part of international climate policy. Until a few years ago, CO_2 extraction processes were also seen as a diversionary tactic by the fossil fuel industries, so that they could continue to conduct their business unimpeded. This is currently changing radically, and an international consensus is emerging that the "net-zero goal" can realistically be achieved only with "carbon dioxide removal" technologies. Nobody would think of banning accident insurance to get people to drive more carefully. To prioritize the moral risk of researching SRM from the outset over the benefits for individuals and society is at the least grossly negligent. At the same time, it is evident that the moral hazard of solar geoengineering can be reduced only if politicians and society understand that it is not—and never can be—a technofix. Therefore, comprehensive information about the limits of geoengineering must be the basis of any discussion around research.

"There are many legitimate questions about solar geoengineering. But in the end, no argument is left that says: You can't do geoengineering." This is how the Italian atmospheric chemist and physicist Daniele Visioni from Cornell University summarizes the current state of research on radiation modification's risks and side effects in a recent interview with me. Like almost all scientists researching solar geoengineering, Visioni also emphasizes: "Before deployment, we must further reduce the uncertainties." The research agenda for this has already been written.

Homework for Research

Let's summarize: Knowing more is always good. Supporters of a de facto research ban on solar radiation modification believe that it is an exception, because among other things, the research creates a "slippery slope" for its deployment. Once the genie has been described, it is out of the bottle. The two main arguments for this are, firstly, the tendency of politicians and societies to be reluctant to approve research investments after implementation, true to the motto that what is technically possible should be attempted. Secondly, according to critics, research on SRM could give the false impression that decarbonization no longer needs to be pursued with full force. Conversely, in open letters and statements, SRM scientists repeatedly emphasize that all counterarguments must be considered. All three classic pillars of climate protection—mitigation, CDR, and adaptation—must be strengthened. However, the potential risks of dimming the sun are not fundamental exclusion criteria per se if the whole set of climate catastrophes happens. Again, we have to think in risk-risk-scenarios. The sole aim of SRM research in the coming years is to create the best possible basis for deciding whether shading the Earth is even an option. The decision-making framework is formed by weighing the risks of using solar radiation modification compared to continuing the current climate protection policy. The best possible basis for political decision-making requires that not only the scientific consequences of SRM for the Earth system, such as potential shifts in weather patterns, are researched but also, in particular, the possible political, social, and economic consequences.[24] This includes the question of the "moral hazard." Open-ended research could then actually conclude that study into solar geoengineering should be stopped because the dangers of rebound effects on the three classic pillars of climate policy are greater than the expected benefits. As explained above, this risk is unlikely if the communicative framework is set correctly to reduce the risk of moral hazard.

Under no circumstances can it be a reason to prevent acquiring more knowledge for informed decisions about using or banning geoengineering. Of course, there must be a firm moratorium on the use of solar geoengineering during the research period. SRM researchers repeatedly emphasize this in unison.[25]

It is perhaps not surprising that the calls for research bans are particularly loud in Europe, which has a strong faction of technology skeptics, particularly from the political and social sciences. It also seems unsurprising to me that researchers and research sponsors in the US are currently providing the most interesting impulses for systematic, interdisciplinary research into geoengineering.[26] The Office of Science and Technology Policy (OSTP)—a presidential advisory committee that always prepares its reports with the broad support of leading scientists in their respective fields—recently made a well-structured and concrete proposal for a research plan with a time horizon of five years.[27] The proposal is based on three key elements.

1. Physical Aspects and Modeling

A broader data basis is necessary to be able to model the effects of solar geoengineering more precisely. To this end, measurements for all SRM-relevant processes and phenomena would have to be intensified—from the ground, in the atmosphere, and above all, with satellites. Among other things, the composition and interaction of gases and aerosols in the stratosphere, the aerosol-cloud interactions, the chemical processes particularly related to ozone, and seasonal characteristics and changes should be observed more closely. With more and improved data, we could feed the existing models better. Above all, the OSTP requires researchers to increase the number and diversity of models and to tailor them to the relevant questions of SRM. This applies both to the physical models that simulate the processes of the Earth system but especially to new and specific models for the global social consequences considering climate justice, social acceptance, risk tolerance toward possible unknowns, and potential negative rebound effects on current climate protection policy ("moral hazard"). The report explicitly calls for physical outdoor experiments to be carried out, to better understand and predict the reflection effect and spread of aerosols, in combination with laboratory experiments. The authors additionally call for monitoring systems to be set up to detect when solar geoengineering is being carried out somewhere in the world—whether openly or secretly.

2. Development for Standardized Deployment Scenarios

When using SRM, which strategy would have what advantages and disadvantages, and for whom? This question can be answered systematically only when researchers compare it with a standardized deployment scenario. According to the OSTP report, such a standard model should, on the one hand, contain carefully designed scenarios intended to achieve specific climate outcomes, such as the mitigation of temperature peaks ("peak shaving") or the targeted cooling of the polar ice caps. On the other hand, operational scenarios that can be implemented without careful planning, international coordination, and against the will of important political actors would also be important. This is particularly relevant for an assessment of possible geopolitical conflicts. As a result, corresponding approaches from game theory and behavioral science would also have to be included in SRM research.

Since the development of geoengineering scenarios is an iterative process, they must be revised regularly based on updated political decisions and new findings from the natural and social sciences.

3. Impact on Human Health and Ecosystems

So far, the impact on public health has played a relatively minor role in SRM research. This is astonishing because global warming has serious negative consequences for human health and well-being. The question is: How many heat deaths can possibly be prevented annually through solar geoengineering?

It will be imperative to clarify in which regions people can benefit particularly greatly from solar geoengineering in terms of their health—i.e., whether, as critics fear, people in the Global South could be disadvantaged in favor of the well-being of people in rich and cooler countries or, conversely, people in climate-vulnerable regions benefit above average.

According to the report, there is also a significant need for research into the possible benefits and risks for ecosystems and biodiversity. Today, we know too little about whether shading tends to stabilize ecosystems compared to greater warming or whether less direct sunlight (and more diffuse light) weakens them compared to scenarios without SRM. To this end, biologists and ecologists, among others, must become more involved in research on geoengineering.

The Decision that Affects Everyone (Literally Everyone)

Impulses from US research and politics are not always received particularly well by the rest of the world. Skepticism about American dominance plays a role, and sometimes resentment of high research budgets and good research conditions, especially at US elite universities. In my conversations with top American researchers on solar geoengineering, I have sensed a high level of awareness of this problem. They would like the rest of the world to get more intensively involved in research and discourse on solar radiation modification. And they wish for more initiatives like the five-year SRM research program launched by the British government in summer 2024, which provides ten million pounds for risk-risk analyses.[28] This does not change the fundamental problem: Anglo-Saxon-dominated research leads to defensive reactions in many parts of the world. This is exactly what could make rational decision-making for the most global of all technologies significantly more difficult.

If humanity dims the sun, this will affect every single person on Earth. Responsible use can happen only on two bedrocks: evidence and acceptance. Therefore, science must first create the initial conditions for a political decision-making process that is as best informed as possible and based on verifiable criteria. Best possible means that as many perspectives as possible from as many smart scientists as possible, from all over the world and from all relevant disciplines, are incorporated into this decision-making basis. Simply put, to model the climatic and ecological, political and economic, and social and health consequences of solar geoengineering, more interdisciplinarity is needed from all parts of the world. Research today is dominated by climate scientists, often with a scientific background. Economists and social scientists are becoming increasingly involved. However, the discourse hardly has a voice from philosophers and ethicists. Communication scientists would also be welcome, especially on the question: How could the "moral hazard" be reduced through factual information?

Science puts forward plausible hypotheses based on available knowledge and verifies or falsifies them with experiments, data, and models that describe reality more precisely. The more scientists discuss a controversial issue from different perspectives, the higher the likelihood that hypotheses will become reliable evidence. The greater the impact of a decision, the more important the evidence is. This is why comprehensive and rigorous peer-review procedures—with a high level of diversity among peers—are crucial regarding the possible use of solar geoengineering. This is the only way to reduce the risk of incorrect or premature use and potentially over-cautious abandonment of a technology that would bring far more benefit to the result than it would cause damage.

Second, solar geoengineering deployment can take place and be carried out with little risk only if it receives global political acceptance. Particularly in the Global South, there is tremendous skepticism toward technology solutions from the North, which is understandable for historical reasons. Europe, for its part, with its traditional distrust of technology, could become a brake that ignores rational arguments. In addition to interdisciplinarity, more internationality is needed. More diversity of models fundamentally creates more predictive power. However, internationally, there is also a cultural component. Ethiopian politicians will trust an SRM model set up and fed by African climate scientists in Nairobi more than one that comes from Harvard, Yale, or Cornell. The nongovernmental organization Degrees, based in London, has been trying to build bridges on this topic for more than ten years. Its declared goal is "a future with experts from every region of the Global South playing a central role in the evaluation and governance of SRM." To this end, Degrees finances research projects in the South with its own research fund and supports climate experts from developing countries in gaining more visibility in the geoengineering community.[29]

Uncertainty is in the nature of the decision. If, when choosing between several options, we know exactly which choice leads to better results according to defined criteria, then in the sense of decision research, it is not a decision but a routine, an algorithm executed by humans (as previously discussed in chapter 1). The political decision for or against the use of solar geoengineering will always have to be made under conditions of uncertainty. One argument from opponents of SRM research is that all modeling is a wasted effort because the actual effects can be measured only during implementation. In this sense, the deployment itself would be the only sensible experiment, an argument that has a kernel of truth. At the same time though, it falls short, at least when it comes to the three realistic technical paths of solar geoengineering. Marine cloud brightening and the thinning of cirrus clouds are reversible within days; stratospheric sulfur injection is reversible within a year at low doses.

The conclusion is that responsible deployment should be carried out cautiously with a slowly increasing dose, with intensive monitoring of desired and undesirable effects. Of course, this requires a comprehensive infrastructure to be able to measure and evaluate the intervention's consequences. In this sense, the careful, step-by-step implementation would actually be an experiment that must be stopped at any time if negative consequences are greater than expected or even unexpected adverse consequences occur. This initially applies at the geophysical level, but of course, also to the economic, social, and political consequences. However, perhaps with the monitoring infrastructure, the world will discover that the intervention works very well—that it brings the desired effects and fewer side effects than feared.

4

LAW:
United Geoengineering Nations

Rogues at the Thermostat

The behavioral economist Massimo Tavoni from the Polytechnic of Milan is conducting interesting laboratory experiments on the fundamental questions of his discipline: Under what conditions do people cooperate? When are they prone to conflict? And what outcome does their behavior have on themselves and others? In 2019, he examined these questions with a view on solar geoengineering by arranging so-called Public Goods Games. Behavioral economists examine, among other things, how actors weigh their individual interests over the common good when making decisions. In the English-speaking world, these experimental setups are also called, somewhat ironically, "Public Good or Bad Games," which, in the case of Tavoni's geoengineering experiment, implies a small foreshadowing of the result.

The empirical economist had one hundred and forty-four students compete against each other in six rounds of a computer game. He gave them a set of desired climatic conditions on Earth, a toolbox of geoengineering techniques, and a toy money budget to finance the chosen measures. The players who best achieved the specified climate goal were paid out their gains in real euros at the end. In summary, the result was that the students were extremely enthusiastic about using "solar geoengineering"—stratospheric aerosol injection. The effect of this was an excessive cooling of the planet as a whole—i.e., a reduction in the public good "climate." As a result, other players—with their own winnings in euros in mind—took measures to artificially reheat the planet so that in the end, chaotic conditions arose in almost all rounds of the game, even leading to "geoengineering wars." This first happened during game rounds with only two players, which corresponds to a bilateral geopolitical conflict. The conflicts got completely out of hand when more than two parties played against each other—i.e., in a multilateral setting. As Tavoni summarizes: "In the end, everybody loses." A "public bad" is a likely outcome, not only when Italian students engage in uncoordinated and unregulated geoengineering in a scientifically designed computer game.[1]

So-called rogue geoengineering (i.e., geoengineering by "rogues" or "lone wolves") and its chaotic consequences for the climate and politics is not only a concern for economists and game theorists from the geoengineering research community. The US military considers the threat of unilateral geoengineering to be so relevant in a multipolar world with clearly intensifying conflicts that it has commissioned its affiliated research projects agency, the Defense Advanced Research Projects Agency (DARPA), to uncover clandestine geoengineering projects. The DARPA scientists have also been tasked with developing strategies for how the US should react in the event of "rogue geoengineering" by an individual country, a community of states, or non-state actors.

Two capabilities play an essential role in detection. First, the team closely monitors the weather and develops algorithms that recognize unusual weather and climate phenomena as possible clues in the jumble of meteorological data. Second, the DARPA unit is looking for indicators that hardware is being developed somewhere that could be used for large-scale geoengineering. This includes, for example, intelligence information from aircraft engines ordered by turbine manufacturers that are suitable for constructing a stratospheric aircraft with a high load capacity, or on test flights being carried out at high altitudes. One of DARPA's worst-case scenarios is finding signs of termination shock because that would mean that geoengineering operations had already been carried out, successfully concealed, and then terminated by rogue actors for whatever reason.

DARPA conducts strategic analytical considerations and game theory experiments on rogue geoengineering in collaboration with the US Naval War College. The strategy analysts' two most important initial questions are: Who would be most interested in cooling their own climate zone or weather region? And who would be capable of geoengineering on a sufficiently large scale? Computer models are used to calculate not only who has suffered but how severely from past extreme weather events caused by climate change and how extensive the damage will be in the future—possible evidence also includes refugee numbers. The models predict potential conflicts and consider how independent a country is economically and militarily. From this, it can be deduced how easily or quickly a possible geoengineering rogue could be stopped through economic sanctions or military intervention. A country with high oil and gas revenues, like Russia, is only moderately impacted by economic sanctions after the attack on Ukraine. This is completely different to a country heavily dependent on agricultural exports, such as Brazil.

The Naval War College estimates that around two dozen countries worldwide would be able to undertake more far-reaching climate interventions if they freed up the resources necessary for military budgets.[2] Dr. Curtis Bell, director of maritime security and governance at the Naval War College, also believes that the probability of a country resorting to geoengineering measures depends mainly on the form of government and the age of political leaders. On the surface, it seems plausible to assume that an autocratic country headed by a "strongman" like Viktor Orbán or Xi Jinping is more likely to engage in radical intervention in the Earth system with international implications. But perhaps this is also a Western error in thinking. Once the consequences of climate change have reached catastrophic proportions, people in a democracy could demand geoengineering from their elected politicians, or politicians could see it as an election campaign opportunity. According to Bell, younger politicians would be more likely to attempt a geoengineering coup than older ones. Although SAI works relatively quickly, the preparations for an intervention are tedious. So, a politician who makes a power play with rogue geoengineering would need the prospect of at least a decade in office to see its engagement through. Based on his models, Bell gave the British magazine *The Economist* a list of countries that the global community should keep on the radar regarding "rogue geoengineering"—Algeria, Australia, Bangladesh, Egypt, India, Indonesia, Libya, Pakistan, Saudi Arabia, and Thailand.

The fact that nongovernmental actors could dim the sun on their own initiative and without consulting other governments, the UN, or international environmental institutions was long treated by geoengineering experts as a purely theoretical thought experiment. The political scientist David G. Victor gave the thought experiment the name "Greenfinger Scenario" in an ironic allusion to the James Bond villain Goldfinger. In this context, the conspiracy theorists from the chemtrail scene and the emotional critics of geoengineering research like to bring Bill Gates's name into play. As described in the last chapter, Gates supports SRM research projects financially and is also part owner of a patent for a device believed to weaken hurricanes by cooling the ocean's surface. Nonetheless, Gates is not a plausible Greenfinger candidate for many reasons. He is extremely rational and a law-abiding citizen of the United States. It may help to remember that he and his ex-wife Melinda have saved about as many lives through philanthropic initiatives as Hitler, Stalin, and Mao *combined* brought to their deaths.[3] But what about the ultra-rich tech titans with obvious narcissistic personality profiles who have multiple citizenships, for example, South African, Canadian, and American? Who, in all brilliant ambivalence, is not only at the forefront of space innovation but also influences US election campaigns and world public opinion with social media, feels called upon to negotiate with Vladimir Putin about war and peace in Ukraine, views human cyborgs as civilizational progress, and personally wants to die on Mars? To date, Elon Musk has not signaled any interest in solar geoengineering. Hopefully it stays that way.

The Make Sunsets helium balloons mentioned earlier have empirically proven that the Silicon Valley mindset and the technical and entrepreneurial skills of tech entrepreneurs are transforming the Greenfinger scenario from a theoretical thought experiment into a serious threat to the global climate and, in the worst conceivable consequence, to world peace. The founders, Luke Iseman and Andrew Song, are provocateurs. Their cleverly initiated PR campaign for geoengineering (and, of course, for themselves) does not pose a serious threat to the climate or the world order. The amounts of sulfur they send into the stratosphere are far too small. Nevertheless, Make Sunsets can be read as an interesting experiment on how solar geoengineering should not be allowed to happen under any circumstances.

First, the two founders claim that they support their campaign scientifically but do not provide any evidence of this. Evidence and transparency are at the core of scientific work, while grandiose claims are the exact opposite. Second, even this small attempt has already led to international political consequences. Make Sunsets released the first balloons in Baja California—i.e., from Mexican territory. After the first press reports, government representatives publicly protested that a US company was carrying out the experiments in Mexico, of all places, and saw this as an "attack on national sovereignty," immediately launching a legislative initiative to ban geoengineering experiments.[4] And third, the start-up's business model is obviously an indicator of how solar radiation modification can put the public good of a stable climate in even greater danger than it already is.

As described earlier, the start-up sells so-called Cooling Credits. Such a market mechanism can make sense for purchasable CO_2 certificates to avoid emissions or even remove CO_2 from the atmosphere through CDR, provided that the certificates trigger additional climate protection measures (which is not always the case). Avoiding and removing greenhouse gases is an expensive and long-term solution. These compensation certificates can provide a sensible economic incentive. For now, however, there is certainly no risk that this stimulus could lead to a dangerous cooling of the planet.

The calculation would be different for cheap solar geoengineering. The more sulfur transported into the stratosphere, the higher the revenue for the SRM entrepreneurs at Make Sunsets. Seen through the lens of an economist, solar geoengineering is far too cheap for a market mechanism. If several companies offered "SRM Cooling Credits" via an online payment, not only would the prices fall, but presumably, chaotic conditions, like those in Massimo Tavoni's behavioral economics games, would soon prevail. Competitive pressure would result in more and more sulfur in the stratosphere. International politics might then have to issue an emergency decree or individual governments might even have to use military means to prevent an out-of-control geoengineering industry from ushering in the next ice age through competition and the pursuit of profit.

In "Design Thinking," the creative, innovative method for problem-solving, there is a rule that should be the basis of every creative process: "State the obvious!" Make the obvious explicit from the start. Regarding the possible use of solar geoengineering, it is obvious that the research, development, and decision-making must not be left to lone states or rich individuals and certainly not to competing commercial providers. A responsible approach can be scientifically informed and consensus-oriented only across all continents, and regulated by the global community. Many geoengineering researchers consider overcoming the political and regulatory challenges to be significantly more difficult than the scientific and technical ones. It is also obvious that international law does not yet provide any guidelines for SRM. There is no institution responsible for regulation and no decision-making process or decision criteria based on which solar radiation modification could be initiated responsibly. However, there are ideas for a comprehensive governance of geoengineering.

The SRM Convention

International climate policy is a tedious and lengthy business. The United Nations Framework Convention on Climate Change, or UNFCCC for short, was signed by one hundred and fifty-four states in Rio de Janeiro in 1992. It came into force two years later. After many climate summits and countless negotiating sessions between experts, almost two decades later, the UNFCCC process produced the Paris Agreement. Those involved celebrated it as a major breakthrough. In many respects, it was proof that the global community could collaborate despite all crises, conflicts, and disagreements by identifying a common interest of great importance. Another ten years later, we are where we are in climate matters. The genuinely committed climate politicians fight with admirable perseverance for every additional inch of progress in climate protection at the annual meeting of the contracting states, the Conference of the Parties (COP). The first obvious conclusion for a framework convention on geoengineering with the participation of as many states as possible is that it will take years to develop. So, the right time to start this process is now. At first glance, the UNFCCC and its most crucial accompanying body, the

Intergovernmental Panel on Climate Change, are the obvious platforms to initiate a structured international consultation process. Almost all states are already involved. There are well-established discussion and negotiation formats, including the participation of nongovernmental organizations. And the UNFCCC is very much driven by a scientific mindset, to which every decision for or against geoengineering must also be subject.

However, there is disagreement among the geoengineering experts, political scientists, and lawyers as to whether the UNFCCC and IPCC are the right forums. Today, at least, many actors still have a too-defensive attitude toward geoengineering. The fear is too strong that the discussion about solar radiation modification could put additional strain on the already difficult negotiations on mitigation and adaptation. Consensus orientation and veto rights when formulating reports and agreements have so far led to the Climate Framework Convention and the IPCC consistently ignoring the issue. Germany is a driving force in nipping any discussion in the bud when, for example, neighboring Switzerland tries to put the issue on the agenda. It is conceivable that this will change in the coming years. However, many geoengineering experts don't want to wait for that and instead rely on other discussion formats or institutions.

Michael Gerrard, director of the Sabin Center for Climate Change Law at Columbia Law School and a long-time environmental lawyer, believes that pressure from informal working groups will have to increase significantly before the UNFCCC and IPCC commit to broaching this topic. János Pásztor, the former United Nations Assistant Secretary General for Climate Change and executive director of the Carnegie Climate Governance Initiative (C2G), is a tireless political and intellectual pioneer in this field. Pásztor has been campaigning internationally for years to finally address the open regulatory issues.[5] The Climate Overshoot Commission, headed by former WTO director-general Pascal Lamy, also lobbies on the issue and advocates for early regulation.[6]

Perhaps in the foreseeable future, the World Economic Forum or the Club of Rome will also become a platform in the pre-political space that discusses solar geoengineering so intensively that the established UN organizations will no longer be able to avoid putting the issue on the climate policy agenda. Perhaps a group of states within the UN, led by an important country such as China or India, could also raise it. The natural candidate for this would be the so-called Group of 77 (G77), the loose association of now more than one hundred and thirty countries of the Global South. The more pronounced the gap becomes between the political desire to achieve the Paris climate goals and the scientific facts about actual emissions, the harder it will be for dogmatically skeptical countries like Germany to block them. They don't have to become a pioneer in the discussion straightaway. But, at the least, they should no longer exercise their explicit and implicit veto rights and finally allow the necessary discourse about the pros and cons of large-scale technical interventions in the climate system, accompanying it critically with productive contributions to a thorough discussion.

A model for successful international coordination and regulation in environmental policy could be the Montreal Protocol for the protection of the ozone layer, which, after a relatively short negotiation process, succeeded in banning the emission of all chemicals containing chlorine and bromine and ultimately created the first agreement to be ratified by all member states of the

United Nations. The Convention on Biological Diversity (CBD) is a repeated reference point of a multilateral, complex legal instrument to contribute to a public environmental good. The same applies to the International Maritime Organization (IMO), which has made shipping much safer and cleaner, especially in the last two decades, through surprisingly strong enforcement mechanisms of its regulations.

Until official formats and possibly new institutions emerge for a geoengineering convention, the informal organizations will continue to work on suggestions as to which the principles—let's call them the "United Nations Framework for Geoengineering" (UNFGE)—should follow. A working group at the University of Oxford laid the foundation for this in 2009 on behalf of the British Parliament. The so-called Oxford Principles of five guidelines for geoengineering research are as follows.

1. Geoengineering to be regulated as a public good
2. Public participation in geoengineering decision-making
3. Disclosure of geoengineering research and open publication of results
4. Independent assessment of impacts
5. Governance before deployment[7]

A decade later, Michael Gerrard presented a comprehensive draft with fourteen well-thought-out individual points in his book *Climate Engineering and the Law*.[8] His list reflects the current state of discussion on a sensible political and legal framework for solar geoengineering and ranges from which type of projects actually fall under SRM regulation to conflict resolution mechanisms due to undesirable side effects.

The political-legal discussion can be condensed into the following seven critical elements of an international convention on solar geoengineering:

1. Definition of the Subject of Regulation

The transition from a large experiment to deployment could be smooth. For the regulation of SRM, it is important to be able to clearly distinguish between research and deployment. Clear limits must be defined, when a possible impact on the atmosphere does not qualify any more as research—and therefore research regulations no longer apply—and the deployment regulations of an SRM convention would kick in. A key aspect here is that this is an experimental project, and its effects and scope are not precisely limited in space and time but will ultimately affect the entire Earth system. This would also apply to the large-scale deployment of marine cloud brightening and cirrus cloud thinning.

2. Institutional Structure and Authorization Process

Every SRM deployment requires approval by a decision-making body of the geoengineering convention, for example, by an "Executive Council," considering the specified decision criteria. The decision criteria are the central result of the deliberation process in the General Assembly of the Convention, supported by a Scientific Advisory Board, analogous to the Intergovernmental Panel on Climate Change of the UNFCCC. An "Ethics and Compliance Committee" monitors the work and decisions of the "Executive Council." A stakeholder engagement forum always gives nongovernmental organizations and private sector actors the opportunity to make their voices heard. The "Executive Council" and its subordinate authorities (see page 119) have the right to condition any approval for implementation.

According to Gerrard's proposal, only member states of the convention can apply for authorization to carry out a geoengineering measure. To do this, they must submit comprehensive scientific documentation with the best possible impact assessment for all options for action, with particularly cautious risk assessments for other countries or regions. To implement the measure, a government can cooperate with private companies, provided they meet the quality criteria also set out in the convention. These approval processes can probably be set up by consensus for smaller, regionally limited geoengineering measures. However, the decisive question remains: How could a decision for or against the widespread use of stratospheric aerosol injection be made with the greatest possible consensus of the global community?

The ideal case would be unanimity, like the Montreal Protocol. However, it cannot be mandatory to ensure the ability to act. The majority rule with veto rights for a few particularly powerful or particularly affected states would be the realpolitik solution, with the permanent states on the United Nations Security Council certainly claiming a special role for themselves. However, the goal must remain to achieve a broad consensus among the vast majority of all states or alliances of states. An approval rate of 75 percent appears to be an ambitious but still achievable target for a positive decision on a cautious approach, as described in chapter 2. However, a three-quarters majority is realistic only if the convention requires exemplary decision-making principles.

3. Decision-Making Principles

Making the decision for or against the use of solar geoengineering is the task of politicians, not scientists. At the same time, a decision of such great consequence must be made free from tactical political considerations as much as possible. This also applies to the question of who carries out the technical measures. Financial considerations or even the hope of profits should not play a role. Solar geoengineering is cheap and could easily be paid for by an international fund.

Uncertainty is part of the nature of all decisions. Otherwise, they are not decisions. Science must create the best possible foundation for decision-making by not compromising on the standards of scientific methods in the case of geoengineering. Models and data must be developed and verified by a wide variety of scientists on all continents. To this end, special transparency and openness to the exchange of data must be ensured. Scientific geoengineering publications should generally be publicly accessible free of charge ("open access"). Not only does this include and involve the scientific communities of poorer countries in the discussion processes, but the entire interested world public would receive a broader basis for information and discussion.

Extensive transparency and openness about research and development processes are fundamentally desirable for all political decisions with major social impact. But especially in the context of solar geoengineering, compromising transparency and scientific evidence would have catastrophic consequences in several respects. It would not only increase the risk of a collective wrong decision but would also endanger the consensus-oriented decision-making process itself. The risk of "rogue geoengineering" increases as climate damage increases.

4. Oversight and Implementation

A global consensus of politicians ideally make the fundamental decision about the extent of an intervention in the Earth system. The same applies to agreeing on who is responsible for the implementation. This is possibly an international consortium of state-controlled entities whose financing must be secured for future decades to exclude the risk of an involuntary termination shock. An international geoengineering oversight authority should determine the details of the deployment, the underlying quality criteria, and applicable safety standards, as well as monitor exact compliance with all requirements. This authority is subordinate to the "Executive Council" anchored in the convention. The International Atomic Energy Agency (IAEA) in Vienna could be a model for this. Let's call it the International Geoengineering Organization (IGEO) for now.

Based on the goals defined in the geoengineering convention, the IGEO should have a high level of decision-making authority in the deployment and follow-up of the measures, for example, regarding the dosage of sulfur aerosols in the stratosphere and the determination of exactly where they are to be released and how. In the case of detailed technical questions and uncertainties arising during the implementation process, disputes or politically motivated decisions can quickly arise, especially in the plenary session of those involved in the convention. Where exactly the lines of authority must be drawn between the plenary session, the "Executive Council," and the oversight authority requires careful and detailed definition by the convention.

One of the most important tasks of the IGEO is to continuously monitor the effects of solar radiation modification using a comprehensive monitoring system and to evaluate the data collected—including that of other institutions, such as the national weather authorities. Therefore, it is about recording the desired or undesirable or unforeseen effects that impact not only the climate but all changes in the entire biosphere. If technical problems or undesirable developments become apparent, the IGEO must be able to intervene quickly and in a legally secure manner. However, when making fundamental decisions such as changing the strategy or possibly canceling the measure, the organization has to make recommendations only to the "Executive Council."

5. Liability and Compensation

Smaller geoengineering projects with exclusively national impacts will continue to be subject to standard national liability regulations. The wheel of liability rules does not have to be reinvented even for errors or accidents in the direct technical implementation with limited impact—for example, a plane crash of a stratospheric aircraft. The international legal system in shipping and aviation provides a good foundation for this. However, a new liability and compensation mechanism is needed in the event of failed or seriously damaged geoengineering measures.

A large fund must be set up as a precautionary measure for such risks to offer compensation where negative effects occur. The historical causes of climate change in the Northern Hemisphere obliges it to pay into the fund primarily, but also those countries that do not achieve their CO_2 savings targets. Conversely, countries that benefit above average from a successful intervention could also subsequently contribute more.

Compensation is made at the request of the geoengineering losers according to criteria negotiated in the convention. Rich countries with high CO_2 emissions in the past can also apply, not just vulnerable ones. Evidence should be provided via the comprehensive monitoring system led by the IGEO. The UNFCCC's Green

Climate Fund is a model that the Geoengineering Compensation Fund (GECF) could be based on. If there is little or no use of the fund, it could either be paid back to the contributing nations—which could initially fill the fund more quickly—or could be used for other mitigation and adaptation measures and to defuse likely growing climate justice conflicts.

6. Penalties and Conflict Resolution Mechanisms

Legal certainty and enforcement deficits are fundamental problems of international law, especially when powerful states do not adhere to agreements to which they have committed themselves. Even an international geoengineering convention will not be able to solve this problem satisfactorily, especially since there is hardly any possibility of preventing a sovereign state from withdrawing from the convention. Nevertheless, member states should define effective penalties if the rules of the treaty are violated. The penalty should be proportional to the economic performance of a country that violates the rules, a framework modeled on a Swiss law of higher fines for wealthy traffic offenders than low-income drivers.

A convention is, by definition, a consensus. However, conflicts are likely if the defined principles are ignored during implementation. Compensating potential geoengineering losers also harbors the potential for conflict. In individual cases, it will always be difficult to determine exactly whether more drought, storms, or extreme rain are causally a result of geoengineering or whether other reasons are possible. Accordingly, the convention should design arbitration procedures from the outset and build a pool of legally competent mediators committed to political neutrality. A secure legal procedure should be designed, or international arbitration should be set up for possible compensation claims.

7. Stability, Continuous Review, and Adjustment

Intervention in the Earth's climate using solar geoengineering will—as shown—have to last for several decades. Only then does it make sense as an interim solution for avoiding and removing greenhouse gases. From this long-term perspective, there is a need for agreed-upon decision-making mechanisms, based on a shared understanding of risk and justice, as well as scientific evidence and transparency, to be firmly anchored in the convention. This is the only way to ensure long-term trust in the agreement and new geo-engineering institutions. At the same time, any geoengineering regulation breaks new political, legal, and technical territory, of all things, in a highly complex and dynamic system, the Earth's climate. This means that the regulatory framework should enable stability, flexibility, and further development. From the outset, the convention must, therefore, be designed to be able to respond to new scientific findings, technical developments, and innovations, as well as to political and social changes, without the regulatory framework as a whole being called into question.

Therefore, the contract text should define regular review and evaluation mechanisms and exchange formats with the relevant stakeholder groups in the global community. It should also standardize processes for improvement suggestions or corrections from scientists, technicians, and lawyers who have expertise in the continuous adaptation of international legal instruments.

Joint Goal and Shared Reality

Let's reactivate the basic rule of design thinking. It is obvious that negotiating an international agreement with major implications and ethical and international legal questions is a lengthy process. It is also obvious that it will take years before solar geoengineering's scientific and technical uncertainties are reduced to a level that enables an informed decision. Scientific findings, in turn, will decisively determine the sensible legal framework.

Therefore, the research community around geoengineering wants scientific research and international legal discussions to take place in parallel and in interaction. Ideally, the growing scientific knowledge about risks and options will be communicated to the regulators on the way to a decision about whether and how to use SRM, and the regulators will approach the question of how the risks can be minimized in an ongoing discussion process. The political optimists among them hope that such a process will be initiated in the coming years under the auspices of the United Nations, perhaps even in a series of negotiations with the UNFCCC and with the support of the IPCC. The seven aforementioned stages for a responsible and stable legal framework could provide a foundation for discussion. A connection to the Framework Convention on Climate Change would also be desirable because it would directly anchor the message institutionally: Solar geoengineering is not a replacement for decarbonization but rather a tactical intermediate step due to time constraints. Solar geoengineering can exist only when closely linked to classic climate policy measures, including accelerated development of carbon removal technologies. But how likely is it that a divided world will agree on an agreed geoengineering framework convention using inclusive diplomacy?

In this context, the optimists point to the history of the Cold War. This demonstrates that people and powers can switch from conflict to cooperation more quickly than seems possible in conflict mode. Treaties from the détente policy of the 1960s to 1980s nurture the hope that the chances of collaboration between previously hostile nations will increase if common interests are identified and established. The cooperation agreements from the Cold War contain elements that are also relevant for an international geoengineering convention. Here are three examples.

- In 1961, the Berlin Wall was built and the Antarctic Treaty came into force, one year before the Cuban Missile Crisis. The US, the Soviet Union, and ten other states from both global political camps, some of whom with specific territorial claims, agreed on the exclusive peaceful use of the continent south of the 60th parallel. The convention stipulated that the focus must be on scientific research, and preserving the unique environment was the highest priority. To date, fifty-seven countries have joined the treaty. It not only encourages research collaboration but also defines mechanisms for mutual inspections.
- Regarding control mechanisms, the Nuclear Non-Proliferation Treaty could provide points of reference under international law. The treaty was initiated in 1968 by the three nuclear powers, the US, the Soviet Union, and Great Britain, and came into force in 1970. When it came to the military and civil use of nuclear power, it created a level of transparency that many considered unthinkable at the height of system competition between East and West. The International Atomic Energy Agency also created a powerful international oversight authority that still functions well today. Despite all the limitations, one thing can be said: With mutual transparency, a level of trust grew between the rival camps that has saved the world from

nuclear war, at least to this day. The regular review conferences that monitor compliance and negotiate sensible further developments could also be a model for geoengineering. Under the Nuclear Non-Proliferation Treaty, this conference takes place every five years. In the case of geoengineering, an annual review would certainly make sense, at least in the beginning.

- Beyond the well-known environmental treaties, there is the development of the World Trade Organization (WTO). There are currently setbacks in world trade due to geopolitical disorder and growing disputes, especially between the US and China. From a longer historical perspective, however, it remains remarkable how well the WTO has succeeded in decoupling common economic interests from geopolitical conflicts. The successful arbitration processes and courts of the WTO can model how possible conflicts resulting from geoengineering can be de-escalated.

In his book *The Avoidable War*, former Australian Prime Minister Kevin Rudd draws up a political action plan on how the foundations for constructive cooperation can be laid again in the current complex world situation.[9] The plan comprises three stages. First, the parties in the conflict—especially the US and China—must determine where their red lines lie in terms of their own national interests. For China, for example, it would be Taiwan. For the US, access to important natural resources or respect for intellectual property. The parties must at least respect these red lines.

Second, the conflicting powers identify areas in which tough economic competition is possible without each other perceiving this as a political threat to national interests. This could include, for example, the development of artificial intelligence or global market access for civilian goods without military significance.

Third, those issues can be addressed together when everyone is on the same page. The state should promote cooperation at all levels. The obvious areas for cooperation despite geostrategic and ideological conflicts include research and medical measures on pandemics, the fight against poverty in the Global South, and, of course, the fight against climate change. Solar geoengineering is clearly one of those third-stage issues.

In the book, Rudd describes with impressive clarity how international disorder can become a world order again. Still, the United Geoengineering Nations, which would have great foresight to create a secure legal framework in which they can make an almost unanimous, science-based decision for the responsible deployment of solar geoengineering, is not a likely scenario. At least that's how Edward Parson, professor of environmental law at UCLA and one of the world's most renowned legal scholars in the field of geoengineering, sees it. Parson first points out that "rogue geoengineering" by state or non-state actors is not only the worst option for dimming the sun but, unfortunately, it also remains a realistic one. He regrets that discussions about solar geoengineering continue to be systematically avoided at the UNFCCC and IPCC. A few years ago, he still hoped that the General Assembly or the General Secretariat of the United Nations would initiate an official world commission on solar geoengineering.[10]

The most important role model for this world commission would be the Brundtland Commission, which, led by the former Norwegian Prime Minister Gro Harlem Brundtland, has strongly influenced the UN's sustainability policy agenda with various reports. All attempts in the direction of solar geoengineering in recent years have been blocked. Switzerland's proposal to set up an expert panel on solar geoengineering at the United Nations Environment Assembly (UNEA) in Nairobi in the spring of 2024 also did not meet with approval. According to the proposal, experts and representatives of international scientific organizations appointed by national governments should discuss the "risks and opportunities" of solar radiation modification in the committee. The US, Saudi Arabia, and Japan, among others, supported the Swiss proposal. A coalition of African states, supported by Colombia, Mexico, Fiji, and Vanuatu, blocked the expert panel because such a commission would undermine "real climate solutions."[11]

The blockers are likely to be aware of a danger that governance experts like Parson have been pointing out for years: Without international inclusive advisory bodies, the likelihood of unregulated release of sulfur aerosols in the legal vacuum above an altitude of twelve miles increases. Only a few geoengineering legal experts believe in a picture-perfect solution under international law. Many consider a path to deployment dominated by power politics to be more likely. A state or a small community of states may present the global community with the alternatives: Either we agree on rules and procedures now, or we cool the Earth on our own.

Of course, interesting behavioral scientific thought experiments can also be built on this realpolitikal constellation, including real threats. The self-empowered "pioneers" who are likely to suffer particularly badly from climate change will assert that they will cool the Earth to the best of their ability based on all available scientific information. The other nations will ask themselves whether this can be trusted or whether "geoengineering rogues" are trying to gain advantages for their own climate region. If the "pioneers" or "rogues" are economically and militarily weak, they may be dissuaded from their plan by the threat of sanctions or military strikes. However, if they are economically and militarily strong, other powerful nations could also abstain from taking tough measures when weighing the risks.

After a few diplomatic moves, the threat from a pioneer or rogue could become an invitation. This invitation would be for as many countries as possible to agree on a set of rules and technology as quickly as possible, for leading scientists from all over the world to be brought on board as consultants, and for a common monitoring system with the necessary data diversity to be installed. Compensation measures, including financing, must be determined before the first planes or balloons or ships take off. Hard climate realpolitik is not the best of all possible starting points for geoengineering regulation. But it could ultimately force a way of dealing with the public good of climate stability in which not everyone loses, like in Tavoni's "public good or bad" game. How could this be achieved? The following chapter develops a possible answer to this in a fictional story following the principles of scenario techniques. Nobody knows the future. But narratives, which paint a plausible picture of a possible future based on present trends, can help us to prepare for difficult situations. I obviously do not know the exact likelihood of each element of my story of a future rogue geoengineer. Nor do I know the precise timeline. But I consider it highly likely that at some point within the next decades someone will seriously confront the world with a plan to dim the sun. Thought experiments based on possible and plausible scenarios—and, ideally, growing ever more accurate with increased research and open dialogue—will be a good exercise to prepare for that moment.

5

SCENARIO:

A Tale from the Year 2040

Daniele Seymour Greenwill was the first at headquarters in Grenada's capital of St. George's. He felt simultaneously tense and exhausted. The previous night in Chicago, first in the CNN television studio, then giving a final fiery speech in front of ten thousand supporters in a basketball arena, he had presented the arguments for his big campaign—"Cooler Planet? You Decide!"—for the thousandth time in twelve months. And yet his pitch "sounded very fresh," his communications director Nancy assured him. Finally, the crypto billionaire flew to the Caribbean island, his second home for over three years. Today, March 4, 2038, was the day in which it would be decided whether the hardships of the last few years, with ever-increasing intensity—particularly the last few months—were worth it.

Greenwill had invested almost one billion US dollars in the first democratic world referendum. "Only the whole world," that's what "DSG," as his supporters called him, had repeatedly declared, "can vote on the most important question in human history." This was the question the cryptographically secured online voting tool posed:

> *Should the International Geoengineering Action Committee (IGAC) stabilize the global average temperature at 2.0°C (3.6°F) above preindustrial levels starting on January 1, 2040?*
>
> *I agree:* ☐
> *I disagree:* ☐
> *I'm undecided:* ☐

In countless interviews and video posts with a total of more than one-hundred billion views, the charismatic tech tycoon explained in English, and fluent French and Spanish, as well as acceptable Arabic, Hindi, and Mandarin, what the IGAC would do in the event of "democratic approval" by the world majority. In five member states of the Alliance of Small Island States—East Timor, Mauritius, Cape Verde, Cuba, and Grenada—hundreds of large, robust helium balloons would rise on hooks, guided on long ropes like kites, into the stratosphere every day and release their loads of two tons of sulfur dioxide. Thanks to high volumes and simple, robust, and reusable technology, it was possible to reduce the price of aerosol injection to two thousand five hundred dollars per ton of SO_2. In the first few years, caution would be exercised, and only five hundred thousand to one million tons of sulfur dioxide would be deployed, resulting in a cooling effect of around 0.1°C (0.18°F). With the careful support of the scientific advisory board, the IGAC would gradually increase the dose starting in 2043, and would set the global average temperature to 2.0°C (3.6°F) above preindustrial levels by 2050 at the latest. The IGAC Foundation would cover the costs of initially one billion dollars and later around fifteen billion dollars annually for at least thirty years.

A provision of three hundred billion dollars was earmarked in a UBS account in Zurich, personally secured by Greenwill's private assets, which currently amounted to $1.1 trillion, of which 70 percent was invested in cryptocurrencies.

Legally, the International Geoengineering Action Committee was on safe ground. Greenwill repeatedly confirmed this, both in writing and in his speeches. No international law prohibited the release of SO_2 into the stratosphere, as stated on the campaign website and supported by reports—available for download—from renowned international lawyers. The balloons would remain in the airspace of the island states from where they would take off. The governments of these countries, in turn, expressly permitted the launches or created a legal basis for them.

The text accompanying the vote also expressly pointed out that an intervention in the form of stratospheric aerosol injection was associated with risks, which would be minimized as much as possible through the implementation principle of cautious advancing and strict scientific observation of the intervention's effects. According to the statutes of the IGAC, the scientific advisory board and ethics committee were also authorized to initiate the discontinuation or tapering of the measure if it emerged during deployment that "the disadvantages of the 'Cooler Planet' campaign clearly outweigh the climatic advantages for the majority of humanity." Meanwhile, all countries worldwide remained urgently called upon to finally radically increase the pace of their decarbonization measures. The faster CO_2 could be removed from the atmosphere, the sooner the "'Cooler Planet' Campaign can end, and the post-fossil age can finally begin."

In the last few weeks before referendum day, Greenwill campaigned, particularly in the large countries of the so-called Global South. In South America, the focus was on Venezuela, Colombia, Ecuador, and Brazil. In Africa, he was especially active in the extended Sahel zone and presented models that predicted cooling would bring more rain rather than less. In Asia, he had campaigned for artificial cooling, especially where many people lived and where he felt support from national politicians. He made eight appearances in India alone, four in Indonesia, and two in the Philippines. The new Prime Minister Vihaan Najangir even received him in New Delhi and wished him "great success" in his project, along with the hope that the "geoengineering veto players" at the UN would finally reconsider their position. Despite years of good business relationships, Greenwill was unable to get a visa for China. But he and his team already had expected they would reach only the people online who were connected to the rest of the world via VPN. Of course, the facts about the climate were a familiar story there, too.

The climate in 2035, 2036, and 2037 was, on average, 1.9°C (3.42°F) above preindustrial levels. Because of the Super El Niño "Hercules," in which Pacific Ocean temperatures rose higher than ever before, the 2°C (3.6°F) plus mark would probably be broken for the first time. The peak of greenhouse gas emissions—reaching a record of forty-four gigatons of carbon dioxide in 2031—actually appeared to have reduced. But in 2037, it was still at forty gigatons, well above the level in the mid-2020s. The concentration of carbon dioxide in the atmosphere, measured at Mount Mauna Loa in Hawaii, was now 492 ppm, after a startling rise of over 70 in the years after the heatwave of 2025.

At around 8:00, Nancy entered the highly air-conditioned conference room at headquarters, a former boutique hotel with a view of the bay and harbor. Shortly afterward, the scientific team led by Chief Geoengineer Sahib arrived, followed by the five Heads of Operations of the individual balloon launches and landing centers. The online voting had begun at 4:00 AM Greenwich Mean Time. It was scheduled to run for exactly twenty-four hours. Greenwill would announce the result to the world at midnight Caribbean time via video message. So far, the numbers looked very good. It was already evening in Asia, and just over 70 percent of votes were in favor of cooling the planet. The same trend emerged in the Arab world and Africa. It was now early afternoon in Europe, and Europeans were voting just as opinion polls commissioned by the IGAC had predicted: 60 percent were in opposition. Throughout the afternoon, the bars for America also climbed as expected: North America, 55 percent in favor, and South America, 65 percent in favor. Greenwill rested that afternoon. The statistical picture had happily solidified when he woke up at 6:00 PM.

The "Cooler Planet" initiative could proceed with global approval of well over 60 percent. And that's what the tech tycoon would do. There was just one small problem. Only nine hundred million global citizens had voted, meaning less than 10 percent of the world's population. In his video message, Greenwill still spoke of "a crystal-clear mandate." He said, as he so often did during the campaign, that global citizens had to take the reins of action into their own hands because international climate policy had so far failed miserably. He also said that it was intolerable that geoengineering was not even discussed by the UNFCCC and IPCC due to the dogmatic obstructive stance of Europe and Russia, despite there being over one million victims every year as a direct result of extreme weather events—and countless more indirectly. The video concluded with the promise: "Based on all available scientific knowledge, we will cool the planet so that almost everyone on every continent will be better off. And no one worse." He didn't mention voter turnout.

On the dot of midnight, Nancy uploaded the video to all major social media channels and sent it, including the press release, to editorial offices all over the world via the huge mailing list. The click rates were gratifyingly good, especially considering the moderate voter turnout. The media response was also gratifying. Would international politics finally pay serious attention to the action? To date, no UN organization had made a single official statement on the referendum. The community of the Small Island States had repeatedly called for support for the IGAC and emphasized the urgency of the intervention with impressive images of evacuation measures from their own particularly low-lying areas. Of the larger nations, India, Pakistan, Bangladesh, Turkey, Indonesia, Saudi Arabia, Iran, and Nigeria had made favorable statements about the initiative, albeit with the caveat that no non-profit organization should intervene in the Earth system with SAI, but that the UN must finally initiate a decision-making and regulatory process.

The US and Chinese governments had ignored the initiative as best as possible and, in response to persistent questions, their government spokespeople announced that, of course, all climate protection measures would continue to be negotiated in the UNFCCC process and that the IPCC would, *of course*, remain the relevant advisory body. On the other hand, Russia had announced that it would take military action against the balloons "if the Russian weather service observes unusual climatic phenomena over the territory and national and economic interests are affected." What was probably meant was that the Northeast Passage through the Arctic Ocean had to remain navigable for seven to nine months.

In the weeks after the referendum, Greenwill and the IGAC repeatedly issued press releases that preparations were now running at full speed. They also posted old videos of the test balloons, which showed how they lifted the large aluminum cartridges containing sulfur dioxide, reminiscent of oil tanks, into the sky. How the containers then automatically opened up, and two tons of compressed SO_2 gas evaporated into the stratosphere within seconds. The initiative's fans diligently shared the videos. But to Greenwill's displeasure, nothing happened at the political level. At least, that's how it looked on the surface because something was happening behind the scenes—in Dhaka.

The Foreign Minister of Bangladesh invited his counterparts from India and Pakistan, including agriculture ministers and national disaster management officers, to a secret three-day consultation on March 11, 2038. The meeting had been planned for a long time. The timing shortly after the "Cooler Planet" referendum was coincidental, but proved to be a good opportunity to reassess the political situation surrounding the use of solar geoengineering. The constellation of the three countries that came together was also no coincidence. India and Pakistan had achieved astonishingly rapid diplomatic rapprochement since the retirement of Narendra Modi as India's prime minister and the rise of the young modernizer, Vihaan Najangir. Since the late 2020s,

Pakistan had earned a reputation as an agent provocateur for solar geoengineering in the IPCC, as the country suffered more and more frequently from increasingly severe floods. Meanwhile, India endured extreme heat waves in the south. Nevertheless, the government in New Delhi had been looking for a reasonably elegant way to postpone its climate protection commitments for one or two decades. The country did not want to guarantee CO_2 neutrality until 2080 at the earliest. This was the only way it could continue its path of amazing economic growth.

Bangladesh was literally up to its neck in seawater. In the last ten years, sea levels had risen by twelve inches, even faster than predicted in the climate models. While more and more islanders from Tuvalu in the South Pacific moved to Australia, the Ministry of Infrastructure in Pakistan could no longer keep up with the construction of dams, even with the major move into the interior of the country. More and more fertile arable land was lost. The foreign ministers' meeting aimed to prepare the basis for a radical push at the next COP climate protection summit in Hyderabad, which would take place the following fall under the chairmanship of Pakistan.

The "advance" of the three was essentially a diplomatic ultimatum: The global community should immediately decide, via a UN resolution, to establish a comprehensive solar geoengineering program by 2042 at the latest, under the scientific leadership of an international council of leading geoengineering researchers with equal numbers based on population. From there, the cooling of the Earth must begin by 2045 at the latest. By then, a fund for the geoengineering losers would have been set up with a starting capital of one hundred billion dollars, financed by the countries with the highest CO_2 burden, true to the motto: Those who pollute, pay! A global geoengineering moratorium must be put in place immediately. Every form of "rogue geoengineering" would be consistently prevented by the community, even by force of arms if necessary. Should the majority of the UNFCCC not agree to the "proposal," Pakistan, India, and Bangladesh would not only financially support initiatives such as "Cooler Planet" through massive purchases of cooling credits but would also carry out their own SAI program with small reusable rockets that had already been developed, primed to launch daily from the new Space Center of the Indian Space Agency (ISRO) in Kanyakumari, on the southern tip of the country.

The consultations between the three delegations in Dhaka went smoothly. By the end of the first day, the ministerial level agreed on all fundamental issues. On the morning of March 13, the time had come: The confidential fifteen-page draft of the communiqué was completed and could be coordinated with the three prime ministers in the coming weeks, so that the result could then be taken on the offensive at the COP meeting. Who then leaked the draft to the *India Times* was never clearly traced. Within twenty-four hours, pretty much every major news portal and TV station in the world had reported on the plan. Everyone was surprised by the geopolitical dynamics of the paper. As expected, the comments were divided: "The states of the South are finally taking the initiative to cool the Earth in a planned manner" versus "The Three Sun Kings are reaching the next level of human megalomania." The majority of skeptical voices came, unsurprisingly, from Europe. However, in Canada and New Zealand, too, one often read about "human hubris," "the final declaration of bankruptcy of climate policy," and an "Icarus project" from South Asia. The Scandinavian media warned of further "polarization of the world, now even on the unanimous topic of climate."

In the Southern Hemisphere, not only the media commented overwhelmingly in the affirmative, but also the majority of state presidents. Dozens of governments said they would join the Pakistan, India, and Bangladesh initiative. Algeria chaired the Group of 77 at the UN in 2038 and convened a special session of the coalition on the sidelines of the General Assembly. At this meeting, a clear opinion quickly emerged in favor of intervention according to the suggestions of the "Asian colleagues." Important states such as Brazil, Nigeria, and South Africa clearly positioned themselves in favor of this and addressed one country in particular in their speeches: China not only had to join but should also ensure that the US and Russia did not block the initiative in the Security Council.

The prime ministers from Pakistan, India, and Bangladesh traveled together to the G77 in New York. Over the last few weeks, the proposals had been fleshed out into three clear points. The required geoengineering would first be carried out in the "peak shaving" design, with the average temperature initially stabilized at a maximum of 2.2°C-plus (3.96°F) for fifteen years. Second, only countries that invest 1 percent of their gross domestic product in climate protection measures could benefit from the compensation fund; poorer countries would be entitled to earmarked loans from the World Bank. Third, the climate sinners of the past should commit to investing at least 0.5 percent of their GDP in carbon dioxide removal. The last point in particular was cleverly chosen. Both China and the US had massively supported CDR companies in recent years, and the technology for CO_2 extraction had made significant progress. However, with only eighty million tons of atmospheric CO_2 extracted worldwide, it has been impossible to slow down climate change measurably or do good business.

India's Prime Minister Najangir gave the closing speech in the special session. It ended with the following: "Dimming the sun is not the wish of 'three sun kings,' as some in Europe call us. Incidentally, decadence is not our problem. Blocking out the sun is the best option we have left. We ask you all to agree to our proposal and take all the necessary steps for the "United Nations Geoengineering Convention" (UNGC) by January 1, 2039." Eighty-nine of one hundred and thirty-four states agreed, including China. The crucial question now was: How would the US react?

Najangir and his two colleagues were optimistic. Before the UN summit, the White House had informally signaled that the United States could envisage agreeing to the UNGC provided that US scientists were strongly represented on the scientific advisory board. It was also clear that the US reserved the right to withdraw from the convention at any time and to prevent geoengineering intervention if it were clearly designed or carried out to the detriment of the national interests of the US. This attitude was no surprise to insiders in Washington politics. In the previous few weeks, three powerful lobbies had campaigned for constructive engagement: the insurance industry, agriculture, and, unsurprisingly, the coal, oil, and gas industries with their dual interests, not just the pressure to curb emissions but also to push their new business models for carbon dioxide removal.

However, what surprised the initiators of the UNGC was how quickly Victoria Baley, former UN ambassador and second-term Republican US president, spoke out. On the airfield, with the presidential helicopter in the background, four words were enough for her: "No veto from us." Now, as everyone in the G77 knew, Europe could continue to act as a doubter. However, it had no relevance. France and Great Britain were open to geoengineering anyway.

Especially in the south and east of Europe, more and more people suffered from heat, storms, and floods. Politically, the European Union was once again divided and would certainly not impose sanctions on the rest of the world against a majority in the UN. It was soon heard from Brussels that "the EU would be involved intensively and constructively in regulatory issues." Of course, that was also meant as a hidden threat. It was intended as such, especially in Brussels. Post-Putin Russia, in turn, could intervene militarily and would stress the prioritizing of Russia's "national interests" regularly. However, Moscow also saw short-term advantages for its fossil industry from a geoengineering convention. The new power clique of India, Pakistan, and Bangladesh also did not want to jeopardize the relatively good relations with many G77 countries.

The realpolitik theorizing had solidified into a new reality at the COP in Hyderabad in the fall. Some older members of the IPCC—scientists and politicians—resigned from their mandates in the summer in protest against the geoengineering initiatives. On the other hand, most UNFCCC officials busily emphasized that geoengineering had to be done now, but in the right way—clouding the sky through intelligent mechanisms, which would accelerate decarbonization rather than slow it down, by emphasizing that SRM is humanity's very last arrow in the quiver to avoid climate tipping points, particularly the melting of West Antarctic's ice-shelf—and that they had been in favor of it for a long time. Of course, solar geoengineering would have to be carried out under the umbrella of the UNFCCC. Given the time pressure, building a new UN institution was impossible, and the UNFCCC was the best way to tie geoengineering to traditional climate protection measures. To the great satisfaction of the "Three Sun Kings," the

COP in Hyderabad already laid down the ten most important cornerstones of the UNFCCC Geoengineering Convention in the final declaration.

- A scientific commission with members from all continents will develop three options for solar geoengineering by January 1, 2043. The aim is to achieve the greatest possible benefit for all regions of the world. The committee makes a recommendation on the design and timing of deployment.
- Until then, a strict ban on all SRM measures applies. The contracting states consistently enforce this on their national territories. To this end, they can request bilateral enforcement assistance from other contracting states, particularly in the case of "rogue geoengineering" by non-state actors.
- The UN General Assembly votes on deployment starting in 2045. The UNFCCC considers approval of 75 percent to be a consensus.
- The financing of the SAI measures must be secured in advance for at least fifty years. The historically heaviest polluters bear the highest proportionate costs.
- The UNFCCC will monitor implementation through a new authority, the International Geoengineering Authority (IGA). The IGA also pools all information on "rogue geoengineering" and its effects on the Earth system so far.
- Regions of the world that suffer climatic disadvantages due to solar radiation modification receive compensation payments from the Solar Geoengineering Risk Fund (SGRF).
- All existing commitments to reduce CO_2 emissions must be adhered to.

- In parallel, the individual countries commit to minimum targets for carbon dioxide removal (CDR). In total, they must amount to at least five gigatons per year by 2050, at least ten gigatons by 2060, and at least twenty gigatons by 2070.
- It is the goal of the international community to slowly reduce solar radiation modification from 2070 onward.
- The average temperature should ideally be at the lower end of the climate target of 1.5°C (2.7°F) of the Paris Agreement by 2080, but by 2090 at the latest.

At the grand reception following the final declaration in Hyderabad, the mood was far from the euphoria following the signing of the Paris Agreement. The conversations had more the character of a collective therapy session in which the participants learned to better deal with their own doubts. A quip from the early days of solar geoengineering often made the rounds: "It's a bad idea whose time has come." There was at least agreement that more countries should have invested in research into solar aerosol injection much earlier. Now, most of the insights came from the US, which, once again presented itself as an "indispensable nation" in world politics, albeit a little more discreetly than in other political areas. The great drought in Africa, the many forest fires in the US and Canada, and the heat wave in southern Europe in the twelve months after the Hyderabad conference probably further accelerated the process. A flood of disaster images no doubt contributed. The campaigns of geoengineering opponents received little attention, and the decision-making in almost all national parliaments and autocratic ruling circles ended in the realization: "It is too risky not to take the risk." In the countries with a skeptical public, rulers repeated the promise like a prayer wheel: "If the sulfur in the stratosphere does not show the desired result, the intervention will be stopped immediately."

The UN invited delegates from every single country to the inaugural meeting of the "Scientific Commission on Geoengineering" in January 2040 in the historic buildings of the League of Nations. In March, the International Geoengineering Authority was founded in Geneva. When filling the position of founding director, there was initially a dispute between the US and India, each wanting to put forward its own candidate. Through the mediation of the Swiss, an agreement was finally reached to go with a Swiss woman with four deputies from the other continents. The most astonishing thing in the following three years before the first launches with Indian hydrogen rockets was how naturally the decision and implementation grew to be received. The UN General Assembly decided in December 2042 with almost 90 percent approval that "as soon as possible, there should be a uniform deployment with a slowly increasing dose of sulfuric acid near the equator." July 2044 turned out to be as soon as possible, as there were ideal conditions at the European space center in Kourou to launch the Indian transport rockets in a slightly southeasterly direction, and to later collect them with a special ship over the Atlantic and refill them with fuel and sulfuric acid tanks.

On July 16, the Secretary-General of the United Nations, the IGA Director, and the Prime Ministers of India, Pakistan, and Bangladesh jointly pressed a large green button, avidly watched by the world press. The IGA director made a comment that was later quoted repeatedly: "Humanity is incredibly bad at thinking and acting with foresight. It needs an acute crisis to become constructive." The phrase "a day of hope" was often used in the speeches. Most climate scientists, however, felt that July 16 was probably more "a day of stomachaches." Meanwhile, the first one hundred rockets whizzed into the sky, and their camera footage sent images of the clouds of fog back to the ground station.

To the small but active anti-geoengineering movement, it still seemed as if the world was deliberately looking the other way, ignoring the major risks. The world disregarded them. Daniele Seymour Greenwill also no longer received any attention despite his continual self-marketing. He wrote an autobiography with the help of a ghostwriter. The publisher scheduled the title's publication date, which was translated into thirty languages, for July 16. The blurb talked about the "father of solar geoengineering," without whom "the technical solution to humanity's greatest challenge would never have come into existence." In terms of content, that wasn't even entirely wrong. Even so, in the first month, fewer than twenty thousand copies of his book were sold worldwide. Greenwill was said to have ordered many of them himself. What the world was waiting for wasn't a book about the past of an emergency operation. It longed for data that showed the procedure was successful and that the fog in the sky had the same effect as the aerosols in the stratosphere did many decades ago after the eruption of Mount Pinatubo. The planet cooled by 0.5°C (0.9°F) in the following twelve months. But nobody actually noticed it at the time.

6

POLITICS:
Responsible Geoengineering

Futures and Target Visions

It is not only military strategists, oil companies, state bankers, and green think tanks that use scenario techniques to imagine possible futures and think them through. It is also a popular game theory in the geoengineering research community. The method was invented in the 1950s by the American physicist Herman Kahn. It was initially used primarily by the military and NASA's space programs. One of the best-known scenarios was produced by the strategy department of the oil and gas company Royal Dutch Shell in the 1990s. It described a protest movement led by young people against the fossil fuel industry in 2020.[1] The Shell scenario did not predict that an introverted student from Stockholm would initiate such a protest movement and that schools would always strike on Fridays. But of course, scenario technology and scenario planning are not about predicting details. However, the story excellently illustrates that climate policy is becoming a generational conflict.

Narrative scenarios are tales about the future, written from the perspective of a future narrator. This is based on assumptions that will affect individuals, societies, or humanity in that future, for example, the increase in global average temperatures by more than 2°C (3.6°F) in 2038 in the previous chapter. The events described in the scenario must be perceived as possible from the perspective of the present. I think it is possible that in the next decade, a person similar to the tycoon Greenwill with a lot of money and sufficiently narcissistic personality traits will try to convince the world of solar geoengineering and also tackle the technical implementation. A plausible event in such a scenario does not necessarily have to be probable from the perspective of the present. To me, it seems more likely that, as the consequences of climate change become increasingly dire, a coalition of states will take the initiative, present themselves as a rational, science-based actor, and through the UN, try to convince as many other states as possible to join the project. I also think the coalition may carry out solar geoengineering even if it does not find a majority in the UN, trusting that the cautious approach method will not lead to military conflicts.

There should be no logical contradictions in plausible scenarios. To be helpful, they must remain understandable and not combine too many possible developments.[2] Military strategists and oil companies may see the greatest value of scenarios in anticipating unfavorable developments and, thus, to act as shrewdly as possible for their own advantage. For political and social questions or, in the case of climate, for human questions, plausible stories of the future have different functions. They should create target visions. It's about developing an idea of what a desirable future could look like under possible framework conditions. When individuals and communities develop a vision of a desirable future, they can work in the present to shape the future in this vein. Isn't this all obvious? Perhaps. However, it is also obvious that the future narrative of the Paris Agreement from 2015 is no longer plausible today. It is obvious that a desirable future for today's young people and their future children can become a horror scenario if global warming continues along the most likely path based on current scientific knowledge. It is obvious that solar geoengineering does not feature in publicly and politically negotiated futures. Such a stance is irresponsible.

It is the task of enlightened, rational-thinking people, who trust in science and believe that the democratic opinion-forming process is the right basis for collective decision-making, to discuss the possibilities of a desirable future openly. By open, I mean considering all options that might make sense from today's perspective, even if they are not an optimal solution or, like solar geoengineering, at best, could be an interim solution. Sometimes, I wonder why solar geoengineering is barely discussed at all. Many of my politically well-informed friends have never heard of the fairly simple option of dimming the sun. Or they ask irritably: "You mean those mirrors in space? That's complete nonsense, isn't it?" I often come away from conversations with climate scientists with the impression that they haven't seriously considered the arguments of David Keith, Anthony Harding, Katherine Ricke, and Edward Parson. When I discuss stratospheric aerosol injection with climate activists, I feel rejection and aggression to a degree I have never experienced in other emotionally charged technology debates, such as genetic engineering or data protection. I can't even choose whether I'd rather be perceived as a hubris-ridden tech solutionist or a morally corrupt oil industry stooge. And to emphasize, even though I explain the analytical drama of the climate situation, it doesn't help, either. To escalate a conversation, all you have to do is say: "You are so caught up in your own narrative that you have lost openness to new information and the ideas of others."

The Tricky Problem of Collective Action

I do not believe in a great force or even a conspiracy that suppresses the discussion of unpleasant topics in liberal democracies. And no, solar geoengineering has nothing to do with chemtrails, either. My impression is that a mechanism is at work here, one that the British philosopher, anthropologist, and social scientist, Steve Rayner, who died in 2020, called "the social construction of ignorance."[3] The concept of "uncomfortable knowledge" plays a central role here. Rayner explained the emergence of collective ignorance, which I have summarized as follows: The world is complex. To remain able to act in this complexity, individuals, institutions, and communities must develop simplified, self-consistent versions of that world. They systematically expunge anything that may be in tension or outright contradiction to this. Uncomfortable knowledge is filtered out using four strategies: denial, dismissal, diversion, or displacement. For Rayner, solar geoengineering was already a prime example of the construction of social ignorance more than ten years ago.

"Uncomfortable knowledge" is particularly problematic in connection with climate change issues, because climate change is one of those "wicked problems" for which uncomfortable knowledge can be of great importance. In the spirit of Steve Rayner, the first conclusion from this non-debate is that collectively ignoring the topic of solar geoengineering uses a typical pattern that is in no way suitable for solving complex problems, namely ignoring uncomfortable knowledge. Simply put, not talking about solar geoengineering is convenient and destructive at the same time.

In the case of the already terribly complicated problem of global warming, what makes matters worse is that collectively ignored knowledge for the purpose of simplification exacerbates the problem in its very essence. This is because climate change is a classic "collective action problem" twice over. First, global warming is the product of collective misconduct because, despite better knowledge of the catastrophic consequences, every year humanity emits more greenhouse gases than the year before. Therefore, the public good of a favorable climate to people, flora, and fauna is subject to the commons' tragedy. Too many people, organizations, and communities value their own short-term benefits from anti-climate behavior over the long-term preservation of the common good of the climate. Second, the solution to the problem (at least in a timely manner) has been failing for decades because of the well-known hurdles that collective action always has to overcome, even if all actors were able to agree on a common goal: Who will lead the way? How much must each actor contribute to maintaining (or rebuilding) the commons? How can rule violations be punished and free-rider effects be prevented? And when can each individual actor expect their present sacrifice to pay off for everyone in the future, including themselves? All these questions need to be clarified, so solving collective action problems takes a lot of time. The climate policy of the last few decades is a textbook example of this.

Of course, an optimistic look at climate policy progress can show that humanity is making progress in the fight against climate change. Common goals have been agreed upon, and almost all countries have committed to the path toward climate neutrality. At the beginning of the century, it seemed likely that the temperature would rise by 6°C (10.8°F) by the year 2100 if humanity did not finally see sense. Today, 3°C (5.4°F) is probably more likely on the current development path. Many companies on all continents, large and small, are investing in green technologies and reducing their ecological footprint. The prices for solar and wind energy are falling rapidly and are cheaper than fossil energy almost everywhere in the world. Peak oil and peak gas are within reach, and a hydrogen economy is slowly taking off. Batteries are getting better and cheaper, and more and more electric cars are on the roads. Many people are eating less meat, thereby reducing greenhouse gas emissions, especially methane. Population growth is slowing sharply, and the world population is clearly approaching its peak. Research into transmuting nuclear waste into fuel for new reactors is progressing well, and technical breakthroughs such as nuclear fusion are possible. Carbon dioxide removal technologies are also evolving. If we produce an abundance of clean energy, we can finally suck gigatons of CO_2 out of the atmosphere. Climate change can then not only be stopped but also reversed by our grandchildren and great-grandchildren. Basically, everything is moving in the right direction, thanks to the many people who fought so hard in politics and NGOs, sustainable companies, and school strikes on Fridays for a good climate policy in the sense of responsible, long-term thinking and action. They have fundamentally solved the biggest problem of collective action with the UNFCCC. Let's celebrate them for that. Unfortunately, this solution needs more time than the world has before tipping points are reached and the Earth's climate slips irreversibly into a new, much warmer equilibrium. And yet, the reason for the slow pace of decarbonization is also obvious.

Climate change is one of those tricky problems in which the costs of solving it are particularly high in the present and the benefits materialize only in the indefinite future. With solar geoengineering, it is the other way around. Viewed through the lens of game theory, dimming the sun is itself a "collective action problem." Chapters 4 and 5 described this in detail. However, the categorical difference to classic climate protection lies in this: The costs at present are very low, and the benefits are immediate. The enlightened geoengineering skeptics may find this to be particularly uncomfortable knowledge, but by solving a much smaller problem of collective action, the time can be bought with which the tricky actual problem can be solved thanks to better will, better knowledge, and better politics and technology.

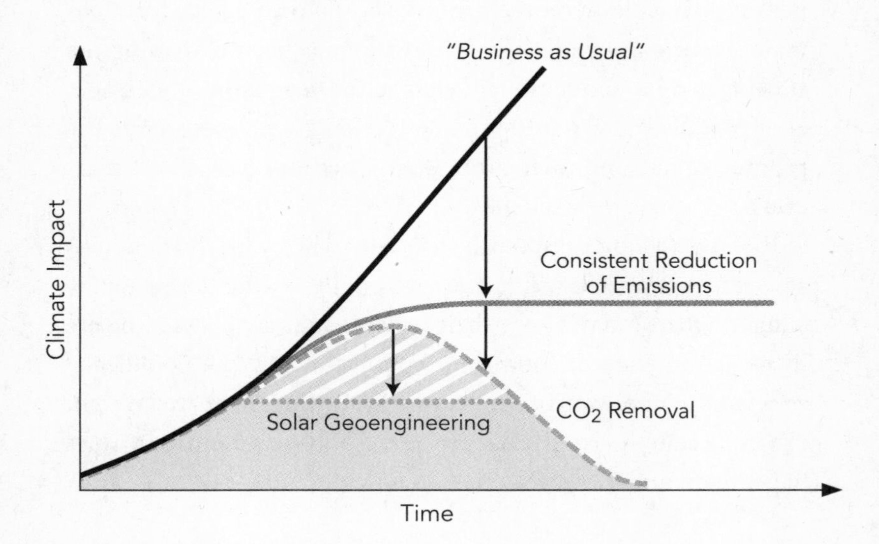

Responsible Geoengineering?

I started this last chapter with the title "Responsible Geoengineering." This is meant as a provocation. Solar geoengineering would be, after all, an act of despair of humankind, which has been acting irresponsibly ever since the causality of CO_2 emissions and global warming has become evident. I, too, hope from the bottom of my heart that, through a stroke of luck, the planet will be spared this intervention. I also hope that all the pessimistic assumptions my argument is based on will prove utterly wrong—that the green transition will finally gain traction much faster than I fear. But as of today, I also yearn for a global climate policy finally developing the ability to consider uncomfortable knowledge and incorporate it into democratic decision-making. This would be an honest climate policy. The current narrative is dishonest. Everyone feels it, but no one says it openly. The debate about the need to decarbonize even faster is now entering the next repetitive loop on the big talk shows, with guests from climate science and climate politics who have been delivering the same messages about climate change for years. These messages are generally correct at their core. However, they exaggerate the progress made and offer no new insights into how the climate catastrophe can be avoided.

The impetus for a responsible climate policy with the option of solar geoengineering will not come from those who dominate the climate debate today. I hope their replacements are people who are interested in science, optimistic about technology, and politically minded. In other words, people who know that there are no right or wrong solutions to wicked problems, only better or worse ways to get them under control.

Technology skeptics would be well advised not to miss the boat on this issue. This is the only way to influence the design of geoengineering if whoever takes action clouds the stratosphere with balloons, airplanes, or rockets. Their contribution could advance the governance of solar geoengineering under the umbrella of the UN environment and climate programs and help think through the security policy aspects from the outset.[4] Responsible skeptics could also work to ensure that all relevant countries and actors agree from the start that SRM should be seen only as a transitional measure—called a "stopgap measure" in UN climate jargon—and not as a solution.[5] Perhaps this will be even more difficult than in my Greenwill scenario. The crypto billionaire and "The Three Sun Kings" would be tangible in world politics because they want to control geoengineering centrally. Another rarely discussed but plausible scenario is that solar geoengineering will be implemented in an uncoordinated and decentralized manner by a wide variety of actors in ten to fifteen years. Maybe a tech billionaire finances guerrilla geoengineering groups in the Global South, which then operate in countries that approve and allow deployment. Maybe start-ups like Make Sunsets have become companies that earn good money with certified cooling credits and, therefore, fly considerably more balloons, possibly with different aerosol precursors rather than sulfur. Perhaps this new generation of climate tech-companies will then receive their growth capital from oil and gas companies. Perhaps smaller states cannot be deterred from sowing large areas of clouds or dissolving cirrus clouds in the summer, regardless of the weather phenomena they trigger for their neighbors, citing their right to climate self-defense. Maybe a seventy-year-old Elon Musk will actually feel called upon to send his SpaceX engineers to conduct tests with moon dust in the extended Earth orbit. In such a scenario with decentralized actors using different geoengineering techniques, not only would the effect be uncontrollable, but worse still, even if the important states then join forces to prevent "rogue geoengineering," they would have to deal with a problem similar to that in combating international drug gangs. With enormous police or even military efforts, only short-term successes could be achieved.[6] That would possibly be the worst of the plausible geoengineering futures.

The idea of dimming the sun, in some ways, disturbs me, because it conflicts so strongly with our image of what humanity should be. It throws us back to what we are: carbon junkies. First, the human species, out of stupidity and greed, screwed up the very climate that made it possible for them to prosper. Then this species, in its self-importance, believed that it could restore the preindustrial climate by tinkering with the thermostat of the Earth system. It's really not flattering to look in the mirror and see humanity's irrational behavior, its callousness, and its hubris. However, allowing this fixation on our worst qualities to guide a decision about geoengineering is also useless.

Humans have carried out geoengineering for more than two hundred years by emitting greenhouse gases. At first, they didn't know what they were doing, but since the 1960s at the latest, they have been heating the Earth in the mode of approving acceptance. The likely outcome today is that this form of human hubris amounts to a temperature increase of at least 2.5°C (4.5°F) and, if the biosphere is unlucky, of over 3°C (5.4°F). No matter how pessimistic one's own view of our relationship with the planet may be, it must not prevent humanity from doing better this time. To be more precise, we must finally follow the science.

Since the beginning, climate science has been more than an attempt to understand the climate better; since the late nineteenth century, a motive has been to manipulate the weather and climatic conditions.[7] We can regret this today from the perspective of scientific ethics. Responsibility toward future generations requires us to turn climate science and geophysics into repair tools against the backdrop of climate change. For this to succeed, the participation of social sciences and economics, engineering and philosophy, political science, and law is also required. I hope that, if we determine we must dim the sun to prevent climate catastrophe, we will have by then found a way to do so with minimal negative impact on the ecosphere. And I hope the decision will be made by a consensus of almost all countries. There's good reason to believe this unpleasant but necessary course of action will motivate people to reduce emissions more quickly and increase carbon dioxide removal more swiftly, like a patient who finally realizes he must live healthier after an emergency operation. In this future, we've avoided the tipping points that would lead to climate catastrophe. The world is neither falling apart politically nor ecologically. After a few decades, we may be able to cease putting sulfur in the sky, and the sun can once again shine as it has for the last one hundred thousand years. The post-fossil age has finally begun.

Notes

Introduction: Sulfur in the Sky

1. Make Sunsets, 2022, makesunsets.com.

2. James Temple, "A startup says it's begun releasing particles into the atmosphere, in an effort to tweak the climate," *MIT Technology Review*, December 24, 2022, technologyreview.com/2022/12/24/1066041/a-startup-says-its-begun-releasing-particles-into-the-atmosphere-in-an-effort-to-tweak-the-climate.

3. "America's defence department is looking for rogue geoengineers," *The Economist*, November 2, 2022, economist.com/science-and-technology/2022/11/02/americas-defence-department-is-looking-for-rogue-geoengineers.

4. René M. van Westen et al., "Physics-based early warning signal shows that AMOC is on tipping course," *ScienceAdvances* 10, no. 6 (2024).

5. Luke Kemp et al., "Climate Endgame: Exploring catastrophic climate change scenarios," *Proceedings of the National Academy of Sciences* 119, no. 34 (2022): e2108146119.

6. The Climate Action Tracker Consortium, "Warming Projections Global Update December 2023," climateactiontracker.org/documents/1187/CAT_2023-12-05_GlobalUpdate_COP28.pdf.

7. "UNEP Emissions Gap Report 2023," unep.org/resources/emissions-gap-report-2023.

8. Copernicus, "The 2023 Annual Climate Summary: Global Climate Highlights 2023," 2024, climate.copernicus.eu/global-climate-highlights-2023.

9. Linda Feldmann, "Newt Gingrich: 8 of the GOP idea man's more unusual ideas," *The Christian Science Monitor*, December 15, 2011, csmonitor.com/USA/Elections/President/2011/1215/Newt-Gingrich-8-of-the-GOP-idea-man-s-more-unusual-ideas/Using-geo-engineering-to-combat-global-warming.

10. Holly Jean Buck et al., "Evaluating the efficacy and equity of environmental stopgap measures," *Nature Sustainability* 3 (2020): 499–504.

11. Rafael Laguna de la Vera and Thomas Ramge, "Darum könnte grüne Energie schon bald kaum noch etwas kosten (That's why green energy could soon cost next to nothing)," in: *Welt am Sonntag*, May 5, 2022, welt.de/debatte/kommentare/plus238360623/Umweltfreundlicher-Strom-Gruene-Energie-wird-eines-Tages-kaum-noch-etwas-kosten.html?cid=socialmedia.twitter.shared.web.

1. Climate: Why We Have to Dim the Sun

1. Steven C. Sherwood, and Matthew Huber. "An adaptability limit to climate change due to heat stress," *Proceedings of the National Academy of Sciences* 107, no. 21 (2010): 9552–55.

2. NASA Science Editorial Team, "Too Hot to Handle: How Climate Change May Make Some Places Too Hot to Live," NASA, last modified March 18, 2024, climate.nasa.gov/explore/ask-nasa-climate/3151/too-hot-to-handle-how-climate-change-may-make-some-places-too-hot-to-live.

3. Ibid.

4. *The Economist*, "Three degrees of global warming is quite plausible and truly disastrous," July 24 2021, economist.com/briefing/2021/07/24/three-degrees-of-global-warming-is-quite-plausible-and-truly-disastrous#.

5. Hanna Metzen, "Wie beschleunigt der Klimawandel das Artensterben? (How is Climate Change Accelerating Species Extinction?)," Press Release of Bielefeld University, November 2 2022, aktuell.uni-bielefeld.de/2022/11/02/wie-beschleunigt-der-klimawandel-das-artensterben; wwf.de/themen-projekte/artensterben.

6. Rachel Warren et al., "The implications of the United Nations Paris Agreement on Climate Change for Globally Significant Biodiversity Areas," *Climatic Change* 147 (2018): 395–409, wwf.org.uk/sites/default/files/2018-03/WWF_Wildlife_in_a_Warming_World.pdf.

7. Petr Chylek et al., "Annual Mean Arctic Amplification 1970–2020: Observed and Simulated by CMIP6 Climate Models," *Geophysical Research Letters* 49, no. 13 (2022).

8. "Coastal Risk Screening Tool," Climate Central, coastal.climatecentral.org.

9. Jonathan L. Bamber et al., "Ice sheet contributions to future sea-level rise from structured expert judgment," *Proceedings of the National Academy of Sciences* 116, no. 23 (2019): 11195–200.

10. Will Steffen et al., "Trajectories of the Earth System in the Anthropocene," *Proceedings of the National Academy of Sciences* 115, no. 33 (2018); 8252–59.

11. "Kippelemente—Großrisiken im Erdsystem: Aktueller Forschungsstand (Tipping Elements—Major Risks in the Earth System)," Potsdam Institute for Climate Impact Research, continuously updated, pik-potsdam.de/de/produkte/infothek/kippelemente.

12. Niklas Boers, "Observation-based early-warning signals for a collapse of the Atlantic Meridional Overturning Circulation," *Nature Climate Change* 11 (2021): 680–88.

13. René M. van Westen et al., "Physics-based early warning signal"; Jonathan Watts, "Atlantic Ocean circulation nearing 'devastating' tipping point, study finds," *The Guardian*, Feburary 9, 2024, theguardian.com/environment/2024/feb/09/atlantic-ocean-circulation-nearing-devastating-tipping-point-study-finds.

14. Tapio Schneider et al., "Possible climate transitions from breakup of stratocumulus decks under greenhouse warming," *Nature Geoscience* 12, no. 3 (2019): 163–67.

15. Nico Wunderling et al., "Interacting tipping elements increase risk of climate domino effects under global warming," *Earth System Dynamics* 12, no. 2 (2021): 601–19.

16. "Überschreiten der Klimaziele könnte das Risiko von Kippeffekten deutlich erhöhen (Overshooting climate targets could significantly increase risk for tipping cascades)," Potsdam Institute for Climate Impact Research, December 22, 2022, pik-potsdam.de/en/news/latest-news/overshooting-climate-targets-could-significantly-increase-risk-for-tipping-cascades.

17. Florian U. Jehn et al., "Betting on the best case: higher end warming is underrepresented in research," *Environmental Research Letters* 16, no. 8 (2021).

18. Ernst Bloch, *Das Prinzip Hoffnung. Werkausgabe: Band 5* (*The Principle of Hope. Work edition: Volume 5*; Suhrkamp Verlag, 1985).

19. Rafael Laguna de la Vera and Thomas Ramge, *On the Brink of Utopia: Reinventing Innovation to Solve the World's Largest Problems* (The MIT Press, 2023).

20. "UNEP Emissions Gap Report 2023," unep.org/resources/emissions-gap-report-2023.

21. "NASA Study: More Greenland Ice Lost Than Previously Estimated," NASA Jet Propulsion Laboratory, California Institute of Technology, January 17, 2024, jpl.nasa.gov/news/nasa-study-more-greenland-ice-lost-than-previously-estimated; Chad A. Greene et al., "Ubiquitous acceleration in Greenland Ice Sheet calving from 1985 to 2022," *Nature* 625 (2024): 523–28.

22. "US Billion-dollar Weather and Climate Disasters," NOAA National Centers for Environmental Information, last modified March 9, 2024, ncei.noaa.gov/access/billions; climate.gov/news-features/blogs/beyond-data/2023-historic-year-us-billion-dollar-weather-and-climate-disasters.

23. "Libya floods: Why damage to Derna was so catastrophic," BBC News, September 14, 2023, bbc.com/news/world-africa-66799518.

24. Adam Morton et al., "Cop28 landmark deal agreed to 'transition away' from fossil fuels," *The Guardian*, December 13, 2023, theguardian.com/environment/2023/dec/13/cop28-landmark-deal-agreed-to-transition-away-from-fossil-fuels.

25. Zeke Hausfather and Pierre Friedlingstein, "Analysis: Growth of Chinese fossil CO2 emissions drives new global record in 2023," *CarbonBrief*, December 5, 2023, carbonbrief.org/analysis-growth-of-chinese-fossil-co2-emissions-drives-new-global-record-in-2023.

26. Gloria Dickie, "Global emissions set to fall only 2% by 2030—UN Report," *Reuters*, November 14, 2023, reuters.com/world/global-emissions-set-fall-only-2-by-2030-un-report-2023-11-14.

27. Pippa Crerar, Fiona Harvey, and Kiran Stacey, "Rishi Sunak announces U-turn on key green targets," *The Guardian*, September 20, 2023, theguardian.com/environment/2023/sep/20/rishi-sunak-confirms-rollback-of-key-green-targets.

28. Helena Horton, "UK net zero policies: what has Sunak scrapped and what do changes mean?," *The Guardian*, September 20, 2023, theguardian.com/politics/2023/sep/20/uk-net-zero-policies-scrapped-what-do-changes-mean.

29. A. Martínez and H. J. Mai, "Trump wants to 'Drill, baby, drill.' What does that mean for climate concerns?," NPR, November 15, 2024, npr.org/2024/11/13/nx-s1-5181963/trump-promises-more-drilling-in-the-u-s-to-boost-fossil-fuel-production.

30. David Keith, *A Case for Climate Engineering*, 99.

31. Naomi Klein, "Capitalism vs. the Climate," *The Nation*, November 9, 2011, thenation.com/article/archive/capitalism-vs-; Naomi Klein, "Geoengineering: Testing the Waters," *The New York Times*, October 27, 2012, nytimes.com/2012/10/28/opinion/sunday/geoengineering-testing-the-waters.html.

32. Jeff Tollefson, "Top climate scientists are sceptical that nations will rein in global warming," *Nature*, November 1, 2021, nature.com/articles/d41586-021-02990-w.

2. Technology: The Geoengineer's Toolbox

1. Oliver Morton, *The Planet Remade: How Geoengineering could Change the World* (Princeton University Press, 2017).

2. Walker Raymond Lee et al., Sunlight Reflection Management Primer, last modified 2024, srmprimer.org/srmprimerwiki.

3. Michael Stamatis et al., "An Assessment of Global Dimming and Brightening during 1984–2018 Using the FORTH Radiative Transfer Model and ISCCP Satellite and MERRA-2 Reanalysis Data," *Atmosphere* 14, no. 8 (2023): 1258.

4. *Climate Change 2021: The Physical Science Basis. Working Group I Contribution to the Sixth Assessment Report of the Intergovernmental Panel on Climate Change*, ed. Valérie Masson-Delmotte et al. (Cambridge University Press, 2021), ipcc.ch/report/ar6/wg1/downloads/report/IPCC_AR6_WGI_FullReport_small.pdf.

5. David Keith and Hadi Dowlatabadi, "A Serious Look at Geoengineering," *Eos, Transactions, American Geophysical Union* 73, no. 27 (1992): 289–93, scholar.harvard.edu/files/davidkeith/files/09_keith_1992_seriouslookatgeoeng_s.pdf.

6. Paul J. Crutzen, "Albedo Enhancement by Stratospheric Sulfur Injections: A Contribution to Resolve a Policy Dilemma?," *Climatic Change* 77 (2006): 211–20.

7. "SCoPEx," Keutsch Group at Harvard, last modified 2024, keutschgroup.com/scopex.

8. "Petition: Support the Indigenous peoples voices call on Harvard to shut down the SCoPEx project," Saami Council, 2021,saamicouncil.net/news-archive/support-the-indigenous-voices-call-on-harvard-to-shut-down-the-scopex-project.

9. Edward Parson, "Solar Geoengineering in the News—Again and Again," *LegalPlanet*, March 15, 2023, legal-planet.org/2023/03/15/solar-geoengineering-in-the-news-again-and-again.

10. Richard Black, "Geoengineering. Risks and benefits," *BBC News*, August 24, 2012, bbc.co.uk/news/science-environment-19371833.

11. Justin McClellan et al., "Geoengineering Cost Analysis: Final Report," Aurora Flight Sciences, July 27, 2011, zerogeoengineering.com/wp-content/uploads/2016/11/AuroraGeoReport-1.pdf.

12. David Keith and Wake Smith, "Solar geoengineering could start soon if it starts small," *MIT Technology Review*, February 5, 2024, technologyreview.com/2024/02/05/1087587/solar-geoengineering-could-start-soon-if-it-starts-small.

13. "Marine Cloud Brightening Program," University of Washington, last modified 2024, atmos.uw.edu/faculty-and-research/marine-cloud-brightening-program.

14. National Research Council, *Climate Intervention: Reflecting Sunlight to Cool Earth* (The National Academies Press, 2015).

15. Paul Voosen, "'We're changing the clouds.' An unforeseen test of geoengineering is fueling record ocean warmth," *Science*, August 2, 2023, science.org/content/article/changing-clouds-unforeseen-test-geoengineering-fueling-record-ocean-warmth.

16. Eli Kintisch, "Technologies," in *Climate engineering and the law: regulation and liability for solar radiation management and carbon dioxide removal*, ed. Michael B. Gerrard and Tracy Hester (Cambridge University Press, 2018), 38.

17. James Temple, "The Growing Case for Geoengineering," *MIT Technology Review*, April 18, 2017, technologyreview.com/2017/04/18/152336/the-growing-case-for-geoengineering.

18. Tim Newcomb, "MIT Scientists Propose 'Space Bubbles' to Deflect Solar Radiation, Ease Climate Change," *Popular Mechanics*, July 7, 2022, popularmechanics.com/space/a40486004/ space-bubbles-climate-change.

19. Benjamin C. Bromley et al., "Dust as a solar shield," *PLOS Climate* 2, no. 2 (2023): e0000133.

20. David Chaum, "Global warming can now be reversed by shade from moon dust placed in space," *AstroCool*, 2022, astrocool.com.

21. Steve Smith et al., "State of Carbon Dioxide Removal—1st Edition," *OSFHOME* (2023).

22. J. Leifeld and L. Menichetti, "The underappreciated potential of peatlands in global climate change mitigation strategies," *Nature Communications* 9, no. 1 (2018).

23. Christopher Schrader, "Umstrittene Tricks, um den Klimawandel aufzuhalten (Controversial Tricks to Stop Climate Change)," *Spektrum der Wissenschaft*, November 24, 2018,spektrum.de/news/koennen-wir-den-klimawandel-mittels-neuer-technologie-aufhalten/1609658.

24. Paul C. Stoy et al., "Opportunities and Trade-offs among BECCS and the Food, Water, Energy, Biodiversity, and Social Systems Nexus at Regional Scales," *BioScience* 68, no. 2 (2018): 100–11.

25. James S. Campbell et al., "Geochemical Negative Emissions Technologies: Part I. Review," *Frontiers in Climate* 4 (2022).

26. Thomas Ramge, host, Sprind Podcast, "#66 Andreas Oschlies 11/6/2023," Bundesagentur für Sprunginnovationen, June 11, 2023, sprind.org/de/podcast/66-andreas-oschlies.

27. Jawad Mustafa et al., "Electrodialysis process for carbon dioxide capture coupled with salinity reduction: A statistical and quantitative investigation," *Desalination* 548, no. 5 (2023): 116263.

28. Philippe Ciais and Christopher Sabine et al., "Carbon and Other Biogeochemical Cycles," *Climate Change 2013: The Physical Science Basis. Contribution of Working Group I to the Fifth Assessment Report of the Intergovernmental Panel on Climate Change*, ed. T. F. Stocker et al. (Cambridge University Press, 2013), 469 and 546–52, ipcc.ch/site/assets/uploads/2018/02/WG1AR5_Chapter06_FINAL.pdf.

3. Research: The Risks and Side Effects

1. Frank Biermann et al., "Open Letter: We Call for an International Non-Use Agreement on Solar Geoengineering," Solar Geoengineering Non-Use Agreement (2022), solargeoeng.org/non-use-agreement/open-letter.

2. Frank Biermann et al., "Solar geoengineering: The case for an international non-use agreement," *WIRES Climate Change* 13, no. 3 (2022): e754.

3. Biermann et al., "Open Letter."

4. "Solar Geoengineering Myths Debunked: Briefing Note #1," Solar Geoengineering Non-Use Agreement, January 2023, solargeoeng.org/wp-content/uploads/2023/02/SGNUA_1_Briefing_note.pdf.

5. Ben Kravitz et al., "Climate model response from the Geoengineering Model Intercomparison Project (GeoMIP)," *Journal of Geophysical Research: Atmospheres* 118, no. 15 (2013): 8320–32.

6. S. Tilmes et al., "Can regional climate engineering save the summer Arctic sea ice?," *Geophysical Research Letters* 41 (2014); acomstaff.acom.ucar.edu/tilmes/documents/grl51275.pdf.

7. Alan Robock et al., "Regional climate responses to geoengineering with tropical and Arctic SO_2 injections," *Journal of Geophysical Research: Atmospheres* 113, no. D16 (2008).

8. Julia Pongratz et al., "Crop yields in a geoengineered climate," *Nature Climate Change* 2, no. 2 (2012): 101–5.

9. Lee, Sunlight Reflection Management Primer.

10. Lina M. Mercado et al., "Impact of changes in diffuse radiation on the global land carbon sink," *Nature* 458, no. 7241 (2009): 1014–17.

11. Daniel M. Murphy, "Effect of Stratospheric Aerosols on Direct Sunlight and Implications for Concentrating Solar Power," *Environmental Science & Technology* 43, no. 8 (2009): 2784–86.

12. "Briefing Note #1," Solar Geoengineering Non-Use Agreement.

13. David Keith, *A Case for Climate Engineering* (The MIT Press, Boston, 2013), 56.

14. Anthony R. Harding et al., "Climate econometric models indicate solar geoengineering would reduce inter-country income inequality," *Nature Communications* 11, no. 1 (2020).

15. Anthony R. Harding et al., "Impact of solar geoengineering on temperature-attributable mortality," Working paper, 2023, rff.org/publications/working-papers/impact-of-solar-geoengineering-on-temperature-attributable-mortality.

16. L. Bengtsson, "Geo-engineering to confine climate change: is it at all feasible?," *Climatic Change* 77 (2006): 229–34. See also: Naomi Klein, *This Changes Everything: Capitalism vs. The Climate* (Simon & Schuster, 2014).

17. Florian Rabitz, "Governing the termination problem in solar radiation management," *Environmental Politics* 28, no. 3 (2019): 502–22.

18. Martin Lukacs, "Trump presidency 'opens door' to planet-hacking geoengineer experiments," *The Guardian*, March 27, 2017, theguardian.com/environment/true-north/2017/mar/27/trump-presidency-opens-door-to-planet-hacking-geoengineer-experiments.

19. Lukacs, "Trump presidency."

20. Gernot Wagner, *Und wenn wir einfach die Sonne verdunkeln? Das riskante Spiel, mit Geoengineering die Klimakrise aufhalten zu wollen* (*Geoengineering: The Gamble*; Oekom Verlag, 2021), 132.

21. Timo Goeschl, Daniel Heyen, and Juan Moreno-Cruz, "The Intergenerational Transfer of Solar Radiation Management Capabilities and Atmospheric Carbon Stocks," *Environmental and Resource Economics* 56, no. 1 (2013): 85–104.

22. Wagner, *Und wenn wir einfach die Sonne verdunkeln?*, 145.

23. Adam Millard-Ball, "The Tuvalu Syndrome," *Climatic Change* 110 (2012): 1047–66

24. Claudia E. Wieners et al., "Solar radiation modification is risky but so is rejecting it: a call for balanced research," *Oxford Open Climate Change* 3, no. 1 (2023). See also: Sarah J. Doherty et al. "An open letter regarding research on reflecting sunlight to reduce the risks of climate change," 2023, climate-intervention-research-letter.org.

25. Parson, "Solar Geoengineering in the News."

26. Joshua B. Horton et al. "Solar geoengineering research programs on national agendas: a comparative analysis of Germany, China, Australia, and the United States," *Climatic Change* 176, no. 4 (2023).

27. Office of Science and Technology Policy (OSTP), "Congressionally Mandated Research Plan and an Initial Research Governance Framework Related to Solar Radiation Modification," Office of Science and Technology Policy, Washington, DC (2023), whitehouse.gov/wp-content/uploads/2023/06/Congressionally-Mandated-Report-on-Solar-Radiation-Modification.pdf. The OSTP report is based on a much more detailed recommendation for SRM research requirements from the National Academies of Sciences, Engineering, and Medicine (NASEM) from 2021. See also: National Academies of Sciences, Engineering, and Medicine, *Reflecting Sunlight: Recommendations for Solar Geoengineering Research and Research Governance* (The National Academies Press, 2021).

28. UK Research and Innovation, "Research programme to model impact of solar radiation management," February 28, 2024, ukri.org/news/research-programme-to-model-impact-of-solar-radiation-management.

29. The Degrees Initiative, 2022.

4. Law: United Geoengineering Nations

1. Anna Lou Abatayo et al., "Solar geoengineering may lead to excessive cooling and high strategic uncertainty," *Proceedings of the National Academy of Sciences* 117, no. 24 (2020): 13393–98.

2. Fort Collins, "America's defence department is looking for rogue geoengineers," *The Economist*, November 2, 2022, economist.com/science-and-technology/2022/11/02/americas-defence-department-is-looking-for-rogue-geoengineers.

3. Oliver Morton, *The Planet Remade: How Geoengineering Could Change the World* (Princeton University Press, 2017).

4. Cassandra Garisson, “Insight: How two weather balloons led Mexico to ban solar geoengineering,” *Reuters*, March 27, 2023, reuters.com/business/environment/how-two-weather-balloons-led-mexico-ban-solar-geoengineering-2023-03-27.

5. Janos Pasztor, “The Need for Governance of Climate Geoengineering,” *Ethics & International Affairs* 31, no. 4 (2017): 419–30. See also: Janos Pasztor, Cynthia Scharf, and Kai-Uwe Barani, “Solar Geoengineering kommt: Zeit, es zu regulieren (Solar Geoengineering Is Coming: Time to Regulate It),” *WirtschaftsWoche*, May 27, 2023, wiwo.de/kai-uwe-barani/29167276.html.

6. Climate Overshoot Commission, “Reducing the Risk of Climate Overshoot,” 2023, overshootcommission.org/report.

7. Memorandum submitted by Tim Kruger et al. (GEO 07), UK Parliament, December 2009, publications.parliament.uk/pa/cm200910/cmselect/cmsctech/221/10011315.htm.

8. Michael B. Gerrard and Tracy Hester, eds., *Climate engineering and the law: regulation and liability for solar radiation management and carbon dioxide removal* (Cambridge University Press, 2018), 331–35.

9. Kevin Rudd, *The Avoidable War: The Dangers of a Catastrophic Conflict between the US and Xi Jinping’s China* (PublicAffairs Books, 2022), 14–15.

10. Edward Parson, "Starting the Dialogue on Climate Engineering Governance: A World Commission," *Centre for International Governance Innovation. Policy Brief: Fixing Climate Governance Series*, no. 8 (2017), cigionline.org/static/documents/documents/Fixing%20Climate%20Governance%20PB%20no8_0.pdf.

11. Joe Lo, "Nations fail to agree to ban or research on solar geoengineering," *Climate Home News*, February 29, 2024, climatechangenews.com/2024/02/29/nations-fail-to-agree-ban-or-research-on-solar-geoengineering-regulations.

6. Politics: Responsible Geoengineering

1. Wagner, *Und wenn wir einfach die Sonne verdunkeln?*, 105.

2. Hannah Kosow and Robert Gaßner, "Methoden der Zukunfts- und Szenarioanalyse: Überblick, Bewertung und Auswahlkriterien. WerkstattBericht Nr. 103 (Methods of future and scenario analysis: Overview, evaluation and selection criteria. Workshop report no. 103)," IZT (Institute for Future Studies and Technology Assessment), 2008, 28.

3. Steve Rayner, "Uncomfortable knowledge: the social construction of ignorance in science and environmental policy discourses," *Economy and Society* 41, no. 1 (2012): 107–25.

4. Oliver Geden and Susanne Dröge, "Vorausschauende Governance für Solares Strahlungsmanagement (Predictive Governance for Solar Radiation Management)," *SWP-Aktuell* 36 (2019), swp-berlin.org/publications/products/aktuell/2019A36_Gdn-Dge.pdf.

5. Buck et al., "Evaluating stopgap measures," 499–504.

6. Wagner, *Und wenn wir einfach die Sonne verdunkeln?*, 125–26.

7. David W. Keith, "Geoengineering the Climate: History and Prospect," *Annual Review of Energy and the Environment* 25 (2000): 245–84.

Further Reading

Abatayo, Anna Lou, et al., "Solar geoengineering may lead to excessive cooling and high strategic uncertainty," *Proceedings of the National Academy of Sciences* 117, no. 24 (2020): 13393–98.

Bamber, Jonathan L., et al., "Ice sheet contributions to future sea-level rise from structured expert judgment," *Proceedings of the National Academy of Sciences* 116, no. 23 (2019): 11195–200.

Bengtsson, L., "Geo-engineering to confine climate change: is it at all feasible?," *Climatic Change* 77 (2006): 229–34. See also: Naomi Klein, *This Changes Everything: Capitalism vs. The Climate* (Simon & Schuster, 2014).

Biermann, Frank, et al., "Open Letter: We Call for an International Non-Use Agreement on Solar Geoengineering," Solar Geoengineering Non-Use Agreement (2022), solargeoeng.org/non-use-agreement/open-letter.

Biermann, Frank, et al., "Solar geoengineering: The case for an international non-use agreement," *WIRES Climate Change* 13, no. 3 (2022): e754.

Black, Richard, "Geoengineering. Risks and benefits," *BBC News*, August 24, 2012, bbc.co.uk/news/science-environment-19371833.

Bloch, Ernst, *Das Prinzip Hoffnung. Werkausgabe: Band 5* (*The Principle of Hope. Work edition: Volume 5*; Suhrkamp Verlag, 1985).

Boers, Niklas, "Observation-based early-warning signals for a collapse of the Atlantic Meridional Overturning Circulation," *Nature Climate Change* 11 (2021): 680–88.

Bromley, Benjamin C., et al., "Dust as a solar shield," *PLOS Climate* 2, no. 2 (2023): e0000133.

Buck, Holly Jean, et al., "Evaluating the efficacy and equity of environmental stopgap measures," *Nature Sustainability* 3 (2020): 499–504.

Campbell, James S., et al., "Geochemical Negative Emissions Technologies: Part I. Review," *Frontiers in Climate* 4 (2022).

Chaum, David, "Global warming can now be reversed by shade from moon dust placed in space," *AstroCool*, 2022, astrocool.com.

Chylek, Petr, et al., "Annual Mean Arctic Amplification 1970–2020: Observed and Simulated by CMIP6 Climate Models," *Geophysical Research Letters* 49, no. 13 (2022).

Ciais, Philippe, and Christopher Sabine, et al., "Carbon and Other Biogeochemical Cycles," *Climate Change 2013: The Physical Science Basis. Contribution of Working Group I to the Fifth Assessment Report of the Intergovernmental Panel on Climate Change*, ed. T. F. Stocker et al. (Cambridge University Press, 2013), 469 and 546–52, ipcc.ch/site/assets/uploads/2018/02/WG1AR5_Chapter06_FINAL.pdf.

Climate Action Tracker Consortium, "Warming Projections Global Update December 2023," climateactiontracker.org/documents/1187/CAT_2023-12-05_GlobalUpdate_COP28.pdf.

Climate Central, "Coastal Risk Screening Tool," coastal.climatecentral.org.

Climate Overshoot Commission, "Reducing the Risks of Climate Overshoot," 2023, overshootcommission.org/report.

Copernicus, "The 2023 Annual Climate Summary: Global Climate Highlights 2023," January 9, 2024, climate.copernicus.eu/global-climate-highlights-2023.

Crutzen, Paul J., "Albedo Enhancement by Stratospheric Sulfur Injections: A Contribution to Resolve a Policy Dilemma?," *Climatic Change* 77 (2006): 211–20.

Dickie, Gloria, "Global emissions set to fall only 2% by 2030—UN Report," *Reuters*, November 14, 2023, reuters.com/world/global-emissions-set-fall-only-2-by-2030-un-report-2023-11-14.

Doherty, Sarah J., et al., "An open letter regarding research on reflecting sunlight to reduce the risks of climate change," 2023, climate-intervention-research-letter.org.

The Economist, "America's defence department is looking for rogue geoengineers," November 2, 2022, economist.com/science-and-technology/2022/11/02/americas-defence-department-is-looking-for-rogue-geoengineers.

The Economist, "Three degrees of global warming is quite plausible and truly disastrous," July 24, 2021, economist.com/briefing/2021/07/24/three-degrees-of-global-warming-is-quite-plausible-and-truly-disastrous#.

Feldmann, Linda, "Newt Gingrich: 8 of the GOP idea man's more unusual ideas," *The Christian Science Monitor*, December 15, 2011, csmonitor.com/USA/Elections/President/2011/1215/Newt-Gingrich-8-of-the-GOP-idea-man-s-more-unusual-ideas/Using-geo-engineering-to-combat-global-warming.

Garisson, Cassandra, "Insight: How two weather balloons led Mexico to ban solar geoengineering," *Reuters*, March 27, 2023, reuters.com/business/environment/how-two-weather-balloons-led-mexico-ban-solar-geoengineering-2023-03-27.

Geden, Oliver, and Susanne Dröge, "Vorausschauende Governance für Solares Strahlungsmanagement (Predictive Governance for Solar Radiation Management)," *SWP-Aktuell* 36 (2019), swp-berlin.org/publications/products/aktuell/2019A36_Gdn-Dge.pdf.

Gerrard, Michael B., and Tracy Hester, eds., *Climate engineering and the law: Regulation and liability for solar radiation management and carbon dioxide removal* (Cambridge University Press, 2019).

Goeschl, Timo, Daniel Heyen, and Juan Moreno-Cruz, "The Intergenerational Transfer of Solar Radiation Management Capabilities and Atmospheric Carbon Stocks," *Environmental and Resource Economics* 56, no. 1 (2013): 85–104.

Greene, Chad A., et al., "Ubiquitous acceleration in Greenland Ice Sheet calving from 1985 to 2022," *Nature* 625 (2024): 523–28.

Harding, Anthony R., et al., "Climate econometric models indicate solar geoengineering would reduce inter-country income inequality," *Nature Communications* 11, no. 1 (2020).

Harding, Anthony R., et al., "Impact of solar geoengineering on temperature-attributable mortality," Working paper, 2023, rff.org/publications/working-papers/impact-of-solar-geoengineering-on-temperature-attributable-mortality.

Hausfather, Zeke, and Pierre Friedlingstein, "Analysis: Growth of Chinese fossil CO_2 emissions drives new global record in 2023," *CarbonBrief*, December 5, 2023, carbonbrief.org/analysis-growth-of-chinese-fossil-co2-emissions-drives-new-global-record-in-2023.

Horton, Helena, "UK net zero policies: what has Sunak scrapped and what do changes mean?," *The Guardian*, September 20, 2023, theguardian.com/politics/2023/sep/20/uk-net-zero-policies-scrapped-what-do-changes-mean.

Horton, Joshua B., et al., "Solar geoengineering research programs on national agendas: a comparative analysis of Germany, China, Australia, and the United States," *Climatic Change* 176, no. 4 (2023).

Jehn, Florian U., et al., "Betting on the best case: higher end warming is underrepresented in research," *Environmental Research Letters* 16, no. 8 (2021).

Keith, David, *A Case for Climate Engineering* (The MIT Press, 2013).

Keith, David, "Geoengineering the Climate: History and Prospect," *Annual Review of Environment and Resources* 25, no.1 (2000): 245–84.

Keith, David, and Hadi Dowlatabadi, "A Serious Look at Geoengineering," *Eos, Transactions, American Geophysical Union* 73, no. 27 (1992): 289–93.

Keith, David, and Wake Smith, "Solar geoengineering could start soon if it starts small," *MIT Technology Review*, February 5, 2024, technologyreview.com/2024/02/05/1087587/solar-geoengineering-could-start-soon-if-it-starts-small.

Kemp, Luke, et al., "Climate Endgame: Exploring catastrophic climate change scenarios," *Proceedings of the National Academy of Science* 119, no. 34 (2022): e2108146119.

"Kippelemente—Großrisiken im Erdsystem: Aktueller Forschungsstand (Tipping Elements—Major Risks in the Earth System)," Potsdam Institute for Climate Impact Research, pik-potsdam.de/de/produkte/infothek/kippelemente.

Klein, Naomi, "Capitalism vs. the Climate," *The Nation*, November 28, 2011, thenation.com/article/archive/capitalism-vs-climate.

Klein, Naomi, "Geoengineering: Testing the Waters," *The New York Times*, October 27, 2012, nytimes.com/2012/10/28/opinion/sunday/geoengineering-testing-the-waters.html.

Klein, Naomi, *This Changes Everything: Capitalism vs. The Climate* (Simon & Schuster, 2014).

Kosow, Hannah, and Robert Gaßner, *Methoden der Zukunfts- und Szenarioanalyse: Überblick, Bewertung und Auswahlkriterien. WerkstattBericht Nr. 103* (*Methods of future and scenario analysis: Overview, evaluation and selection criteria. Workshop report no. 103*), IZT (Institute for Future Studies and Technology Assessment), 2008.

Kravitz, Ben, et al., "Climate model response from the Geoengineering Model Intercomparison Project (GeoMIP)," *Journal of Geophysical Research: Atmospheres* 118, no. 15 (2013): 8320–32.

Laguna de la Vera, Rafael, and Thomas Ramge, "Darum könnte grüne Energie schon bald kaum noch etwas kosten (That's why green energy could soon cost next to nothing)," in: *Welt am Sonntag*, May 5, 2022, welt.de/debatte/kommentare/plus238360623/Umweltfreundlicher-Strom-Gruene-Energie-wird-eines-Tages-kaum-noch-etwas-kosten.html?cid=socialmedia.twitter.shared.web.

Laguna de la Vera, Rafael, and Thomas Ramge, *On the Brink of Utopia—Reinventing Innovation to Solve the World's Largest Problems* (The MIT Press, 2023).

Leifeld, J. and L. Menichetti, "The underappreciated potential of peatlands in global climate change mitigation strategies," *Nature Communications* 9, no. 1 (2018).

Lee, Walker Raymond, et al., Sunlight Reflection Management Primer, last modified 2024, srmprimer.org/srmprimerwiki.

Lo, Joe, "Nations fail to agree ban or research on solar geoengineering," *Climate Home News*, February 29, 2024, climatechangenews.com/2024/02/29/nations-fail-to-agree-ban-or-research-on-solar-geoengineering-regulations.

Lukacs, Martin, "Trump presidency 'opens door' to planet-hacking geoengineer experiments," *The Guardian*, March 27, 2017, theguardian.com/environment/true-north/2017/mar/27/trump-presidency-opens-door-to-planet-hacking-geoengineer-experiments.

"Marine Cloud Brightening Program," University of Washington, Department of Atmospheric and Climate Science, last modified 2024, atmos.uw.edu/faculty-and-research/marine-cloud-brightening-program.

Memorandum submitted by Tim Kruger et al. (GEO 07), UK Parliament, December 2009, publications.parliament.uk/pa/cm200910/cmselect/cmsctech/221/10011315.htm.

Mercado, Lina M., et al., "Impact of changes in diffuse radiation on the global land carbon sink," *Nature* 458, no. 7241 (2009): 1014–17.

Metzen, Hanna, "Wie beschleunigt der Klimawandel das Artensterben? (How Is Climate Change Accelerating Species Extinction?)," Press Release of Universität Bielefeld, November 2, 2022, aktuell.uni-bielefeld.de/2022/11/02/wie-beschleunigt-der-klimawandel-das-artensterben.

Millard-Ball, Adam, "The Tuvalu Syndrome," *Climatic Change* 110 (2012): 1047–66.

Morton, Adam, et al., "Cop28 landmark deal agreed to 'transition away' from fossil fuels," *The Guardian*, December 13, 2023, theguardian.com/environment/2023/dec/13/cop28-landmark-deal-agreed-to-transition-away-from-fossil-fuels.

Morton, Oliver, *The Planet Remade—How Geoengineering Could Change the World* (Princeton University Press, 2017).

Murphy, Daniel M., "Effect of Stratospheric Aerosols on Direct Sunlight and Implications for Concentrating Solar Power," *Environmental Science & Technology* 43, no. 8 (2009): 2784–86.

Mustafa, Jawad, et al., "Electrodialysis process for carbon dioxide capture coupled with salinity reduction: A statistical and quantitative investigation," *Desalination* 548, no. 5 (2023): 116263.

NASA Science Editorial Team, "Too Hot to Handle: How Climate Change May Make Some Places Too Hot to Live," NASA, last modified March 18, 2024, climate.nasa.gov/explore/ask-nasa-climate/3151/too-hot-to-handle-how-climate-change-may-make-some-places-too-hot-to-live.

National Academies of Sciences, Engineering, and Medicine, *Reflecting Sunlight: Recommendations for Solar Geoengineering Research and Research Governance* (The National Academies Press, 2021).

National Research Council, *Climate Intervention: Reflecting Sunlight to Cool Earth* (The National Academies Press, 2015).

Newcomb, Tim, "MIT Scientists Propose 'Space Bubbles' to Deflect Solar Radiation, Ease Climate Change," *Popular Mechanics*, July 7, 2022, popularmechanics.com/space/a40486004/space-bubbles-climate-change.

NOAA National Centers for Environmental Information (NCEI), "US Billion-Dollar Weather and Climate Disasters," last modified March 9, 2024, ncei.noaa.gov/access/billions; climate.gov/news-features/blogs/beyond-data/2023-historic-year-us-billion-dollar-weather-and-climate-disasters.

Office of Science and Technology Policy (OSTP), "Congressionally Mandated Research Plan and an Initial Research Governance Framework Related to Solar Radiation Modification," Office of Science and Technology Policy, Washington, DC, 2023, whitehouse.gov/wp-content/uploads/2023/06/Congressionally-Mandated-Report-on-Solar-Radiation-Modification.pdf.

Parson, Edward, "Solar Geoengineering in the News—Again and Again," *LegalPlanet*, March 15, 2023, legal-planet.org/2023/03/15/solar-geoengineering-in-the-news-again-and-again.

Parson, Edward, "Starting the Dialogue on Climate Engineering Governance: A World Commission," *Centre for International Governance Innovation, Policy Brief: Fixing Climate Governance Series* no. 8, August 2017, cigionline.org/static/documents/documents/Fixing%20Climate%20Governance%20PB%20no8_0.pdf.

Pasztor, Janos, "The Need for Governance of Climate Geoengineering," *Ethics & International Affairs* 31, no. 4 (2017): 419–30.

Pasztor, Janos, Cynthia Scharf, and Kai-Uwe Barani, "Solar Geoengineering kommt: Zeit, es zu regulieren (Solar Geoengineering Is Coming: Time to Regulate It)," *WirtschaftsWoche*, March 27, 2023, wiwo.de/my/politik/ausland/klimawandel-solar-geoengineering-kommt-zeit-es-zu-regulieren/29167250.html.

"Petition: Support the Indigenous peoples voices call on Harvard to shut down the SCoPEx project," Saami Council, 2021, saamicouncil.net/news-archive/support-the-indigenous-voices-call-on-harvard-to-shut-down-the-scopex-project.

Pongratz, Julia, et al., "Crop yields in a geoengineered climate," *Nature Climate Change* 2 (2012): 101–5.

Rabitz, Florian, "Governing the termination problem in solar radiation management," *Environmental Politics* 28, no. 3 (2019): 502–22.

Rayner, Steve, "Uncomfortable knowledge: the social construction of ignorance in science and environmental policy discourses," *Economy and Society* 41, no. 1 (2012): 107–25.

Robock, Alan, et al., "Regional climate responses to geoengineering with tropical and Arctic SO_2 injections," *Journal of Geophysical Research: Atmospheres* 113, no. D16 (2008).

Rudd, Kevin, *The Avoidable War: The Dangers of a Catastrophic Conflict between the US and Xi Jinping's China* (PublicAffairs Books, 2022).

Schneider, Tapio, et al., "Possible climate transitions from breakup of stratocumulus decks under greenhouse warming," *Nature Geoscience* 12, no. 3 (2019): 163–7.

Schrader, Christopher, "Umstrittene Tricks, um den Klimawandel aufzuhalten (Controversial Tricks to Stop Climate Change)," *Spektrum der Wissenschaft*, November 24, 2018,spektrum.de/news/koennen-wir-den-klimawandel-mittels-neuer-technologie-aufhalten/1609658.

Sherwood, Steven C., and Matthew Huber, "An adaptability limit to climate change due to heat stress," *Proceedings of the National Academy of Sciences* 107, no. 21 (2010): 9552–55.

Smith, Steve, et al., "State of Carbon Dioxide Removal—1st Edition," *OSFHOME* (2023).

Stamatis, Michael, et al., "An Assessment of Global Dimming and Brightening during 1984–2018 Using the FORTH Radiative Transfer Model and ISCCP Satellite and MERRA-2 Reanalysis Data," *Atmosphere* 14, no. 8 (2023): 1258.

Steffen, Will, et al., "Trajectories of the Earth System in the Anthropocene," *Proceedings of the National Academy of Sciences* 115, no. 33 (2018): 8252–59.

Stoy, Paul C., et al., "Opportunities and Trade-offs among BECCS and the Food, Water, Energy, Biodiversity, and Social Systems Nexus at Regional Scales," *BioScience* 68, no. 2 (2018): 100–11.

Temple, James, "A startup says it's begun releasing particles into the atmosphere, in an effort to tweak the climate," *MIT Technology Review*, December 24, 2022, technologyreview.com/2022/12/24/1066041/a-startup-says-its-begun-releasing-particles-into-the-atmosphere-in-an-effort-to-tweak-the-climate.

Temple, James, "The Growing Case for Geoengineering," *MIT Technology Review*, April 18, 2017, technologyreview.com/2017/04/18/152336/the-growing-case-for-geoengineering.

Tilmes, S., et al., "Can regional climate engineering save the summer Arctic sea ice?," *Geophysical Research Letters* 41 (2014).

Tollefson, Jeff, "Top climate scientists are sceptical that nations will rein in global warming," *Nature* 599, November 1, 2021, nature.com/articles/d41586-021-02990-w.

"Überschreiten der Klimaziele könnte das Risiko von Kippeffekten deutlich erhöhen (Overshooting climate targets could significantly increase risk for tipping cascades)," Potsdam Institute for Climate Impact Research, December 22, 2022, pik-potsdam.de/en/news/latest-news/overshooting-climate-targets-could-significantly-increase-risk-for-tipping-cascades.

"UNEP Emissions Gap Report 2023," unep.org/resources/emissions-gap-report-2023.

van der Bruggen, Bart, "Electrodialysis process for carbon dioxide capture coupled with salinity reduction: A statistical and quantitative investigation." *Desalination* 548, no. 5 (2023): 116263.

van Westen, René M., et al., "Physics-based early warning signal shows that AMOC is on tipping course," *ScienceAdvances* 10, no. 6 (2024).

Voosen, Paul, "'We're changing the clouds.' An unforeseen test of geoengineering is fueling record ocean warmth," *Science*, August 2, 2023, science.org/content/article/changing-clouds-unforeseen-test-geoengineering-fueling-record-ocean-warmth.

Wagner, Gernot, *Und wenn wir einfach die Sonne verdunkeln? Das riskante Spiel, mit Geoengineering die Klimakrise aufhalten zu wollen* (*Geoengineering: The Gamble*; Oekom Verlag, 2021).

Warren, Rachel, et al., "The implications of the United Nations Paris Agreement on Climate Change for Globally Significant Biodiversity Areas," *Climatic Change* 147 (2018): 395–409, wwf.org.uk/sites/default/files/2018-03/WWF_Wildlife_in_a_Warming_World.pdf.

Watts, Jonathan, "Atlantic Ocean circulation nearing 'devastating' tipping point, study finds," *The Guardian*, February 9, 2024, theguardian.com/environment/2024/feb/09/atlantic-ocean-circulation-nearing-devastating-tipping-point-study-finds.

Wieners, Claudia E., et al., "Solar radiation modification is risky, but so is rejecting it: a call for balanced research," *Oxford Open Climate Change* 3 no. 1 (2023).

Wunderling, Nico, et al., "Interacting tipping elements increase risk of climate domino effects under global warming," *Earth System Dynamics* 12, no. 2 (2021): 601–19.

Acknowledgments

Dr. Annette Anton helped to launch this book. I thank her very much for immediately recognizing the importance of a controversial topic and for helping me to refine my thoughts so that they are (hopefully) clear but (hopefully) not in danger of becoming ideological or even polemical.

Despite a full calendar, several outstanding scientists supported me in my research with long conversations. I thank Prof. Dr. David Keith, Prof. Dr. Edward Parson, Prof. Dr. Douglas MacMartin, Dr. Daniele Visioni, Dr. Oliver Geden, Prof. Dr. Sascha Friesike, Dr. Max Neufeind, Prof. Dr. Stephan Rammler, Prof. Dr. Stephan Pfahl, Dr. Fabian Hoffmann, and Dr. Stefan Schäfer. These discussions were extremely helpful to me, not only because they gave me access to background information that cannot be found in scientific papers and newspaper texts, but they also helped me question my own, often hasty, conclusions and to reset the book's focus.

I owe great thanks to Matthew Lore and Nick Cizek for making this US edition happen and helping me to bring a global topic to a global audience.

As always, I would like to thank my family from the bottom of my heart for stoically enduring my moods during the writing process.

Index

NOTE: Page references in *italics* refer to illustrations.

About the Author

Dr. Thomas Ramge thinks and writes at the crossroads of technology and economics, sustainability and society. He has published more than twenty nonfiction books, selling more than two million copies worldwide, including *Who's Afraid of AI?: Fear and Promise in the Age of Thinking Machines*, *On the Brink of Utopia: Reinventing Innovation to Solve the World's Largest Problems*, *Reinventing Capitalism in the Age of Big Data*, coauthored with Viktor Mayer-Schönberger, and *The Global Economy as You've Never Seen It*, written with Jan Schwochow. His essays and articles appear in *The Economist*, *Harvard Business Review*, *MIT Sloan Management Review*, and *Foreign Affairs*. He holds a PhD in sociology of technology (on AI-assisted decision-making) and is an Associated Researcher at the Einstein Center Digital Future. He also hosts the podcast SPRIND in collaboration with the Federal Agency for Disruptive Innovations of Germany. Thomas's work has been translated into twenty languages and has received numerous publishing awards, including the German Essay Prize 2022, the Axiom Business Book Award 2019 (Gold Medal, Economics), PWC Best Business Book of the Year on Technology and Innovation 2018, getAbstract International Book Prize 2018, the German Business Book Award, and the ADC Award for excellence in design and craft. He lives in Berlin with his wife and son.